CURRENT AFRICAN ISSUES 54

I0122941

The Oil Industry in Uganda; A Blessing in Disguise or an all Too Familiar Curse?

The 2012 Claude Ake Memorial Lecture

Pamela K. Mbabazi

NORDISKA AFRIKAINSTITUTET, UPPSALA 2013

INDEXING TERMS:
Uganda
Petroleum industry
Natural resources
Economic growth
Government policy
Development strategy

ISSN 0280-2171
ISBN 978-91-7106-740-1
© The author, The Nordic Africa Institute, Uppsala University and
the Department of Peace and Conflict Research
Production: Byrå4
Print on demand, Lightning Source UK Ltd.

Contents

Editor's Preface

This is No. 7 in the CAMP series, the series of research reports presenting the text version of the Claude Ake Memorial Lectures given at end of a longer research stay at Uppsala University and the Nordic Africa Institute by the annual holder of the Claude Ake Visiting Chair.

The 2012 Claude Ake Memorial Lecture was delivered by Dr. Pamela Mbabazi in April 2013. She is an Associate Professor of Development Studies at Mbarara University of Science & Technology in Uganda and currently the Deputy Vice Chancellor. She has been lecturing and doing research in development planning, rural development, political economy and conflict studies over the past 16 years. Her research interests include the political economy of oil as well as land governance issues. She arrived in Uppsala, Sweden in November 2012 and contributed to the research seminars of the two host institutions, the Department of Peace and Conflict Research at the university and the Nordic Africa Institute during the following months, while work on this particular piece of research.

As the issue of oil has drawn global concern among social scientists Dr. Mbabazi selected to work on this topic while in Uppsala. Given the recent discoveries of oil in Uganda and the government's optimistic plans for its utilization, this was a real opportunity for Dr. Mbabazi to explore the issue. Her final presentation drew considerable attention. The official discussant as well as the audience engaged in a lively conversation on this topic. Thus, it is with great satisfaction it is now possible to present this addition to the CAMP series. We are grateful to Dr. Mbabazi for her work and believe it will continue to generate interest among scholars, teachers and policy makers. As is customary to note, this publication constitutes the work of the author and does not necessarily reflect the views of the host institutions or the editor.

Uppsala, Sweden, September 2013

Peter Wallensteen
Professor, CAMP Series Editor

Abstract

The resource curse, as set forth by Richard Auty (1993), indicates that countries highly endowed with natural resources can experience slow economic growth, resulting in a host of other problems, most notably conflict. Classical economics suggests that if a country has a significant amount of a resource in high demand globally than that country should prosper. The resource curse, otherwise known as the paradox of plenty, suggests otherwise. As Uganda limps closer and closer to full out commercial production, the state has much work to do to become Africa's first oil success story. Although much of the literature regarding oil globally, as well as in Uganda, paints a rather pessimistic picture, I provide a set of alternatives, looking at oil as an opportunity rather than a curse.

While oil is still in its infancy, many in Uganda have already forecasted how it will play out over the next 50 years. While some are quick to point out the flaws and potential problem areas, I try to provide a more balanced approach, recognizing both the issue areas as well as the opportunities presented. Uganda has just celebrated its 50th anniversary as an independent nation. What is certain is that over the next 5 decades, oil will play a significant role in Uganda's development. I argue that by heeding the suggestions made in my paper, the government and key decision makers can help set Uganda on the right path to becoming Africa's first oil success story.

For comments & questions contact the author at; pkmbabazi@infocom.co.ug

1. Introduction

In 2008 Uganda's President Yoweri Museveni announced to the world that Uganda would soon become a global oil player. While the president's jubilation was clear, other stakeholders, such as the citizenry and the general media seemed to let out a rather large sigh. This is due in large part to two factors, the failure of other African nations to properly use oil revenues and the high level of corruption currently limiting the nation. Although much of the literature regarding oil globally, as well as in Uganda, paints a rather pessimistic picture, this paper tries to provide a set of alternatives, looking at oil as an opportunity rather than a curse.

The information presented in this paper was collected via a critical review of published literature, government documents and press reports. These were augmented with primary data collected via interviews with leading politicians, top bureaucrats in the Ministry of Energy and Mineral Development, and leading researchers and journalists in Uganda.

The paper attempts to answer five key questions; namely; what are the general opinions/views of the oil industry in Uganda? Why are the majority of those views critical? Is this criticism justified? What can be done to avert this criticism, or turn it into optimism and lastly, what can Uganda learn from a relatively successful African country like Botswana, which has managed to use a natural resource to transform itself into a middle income country.

The paper is divided into five key sections, which try to address the key questions posed above. In the first section I try to explain the resource curse and provide the main tenants of the theory. I also present the rather obvious argument forwarded by Ross (2002) who stated that not all resource endowed states are doomed to fail as evidenced by the success of Botswana and Norway.

In the second section, I explore what the media, both domestically and internationally, are saying about the industry. This section tries to analyse the literature and provide context to the situation on the ground. The section is designed to give the reader a brief introduction to oil in Uganda, as well as a detailed understanding of the sentiments surrounding the resource. Throughout the section, I explore how Uganda has been able to make progress since 1986 when President Museveni came to power. By discussing examples such as Uganda's ability to combat HIV/AIDS; its ability to create national peace and stability (following the defeat of the LRA and all other rebel groups) and its ability to strengthen some public institutions, I argue that we cannot immediately write off Uganda. I note how policies such as the construction of a domestic oil refinery, as well as using reserves to satisfy the domestic market before exporting the commodity demonstrate some of the good policies Uganda is making and some lessons the government appears to have learnt from past failures. Nevertheless, much as Uganda has made progress in some areas, I do note some red flags notably the high level of corruption and high poverty rates which are causes for concern.

In the third section I begin to explore the policies in place that cover the emerging oil industry. I particularly highlight the recently passed controversial oil bill which raised a lot of eyebrows. By introducing government policy towards oil I intend to explain why much of the coverage of the resource has been critical.

In the fourth section of this paper I explore what explains Botswana's unique development trajectory as it is considered to be Africa's only resource success. In this section, I try to highlight what explains Botswana's success and the lessons other countries can learn from this African Success story.

In the fifth and final section I try to analyse whether or not the criticism has been justified, as well as present the government with some opportunities to learn from. In this section, I begin to address what can be done to turn criticism into optimism. I do this by providing some policy options that will set Uganda on a path more in line with Botswana, rather than Nigeria, Sierra Leone, or other well-documented resource failures.

Uganda is undoubtedly at a crossroads. What we do with our oil will likely have a dramatic effect on the long term future of our country. For Uganda to be successful, this must entail a combination of state-building projects (institutions and infrastructure) and creating an enabling environment that will transform the economy from an enclave extractive economy to providing employment opportunities for the larger section of Uganda's ever increasing population. While the oil industry is still in its infancy, many in Uganda have already forecasted how it will play out over the next 50 years. While many are quick to point out the flaws and potential problem areas, I provide a more balanced approach, recognizing both the issue areas as well as the opportunities presented. Uganda has just celebrated its 50th anniversary as an independent nation and although there are a lot of uncertainties about the future of the country especially with regard to increasing corruption and high levels of unemployment, what is very certain is that over the next 5 decades, oil will play a significant role in Uganda's development. By learning from the mistakes made by others and heeding the warnings presented by available literature, I argue that Uganda can utilise its oil to promote development. I believe that by heeding the suggestions made in this paper the government and key decision makers can help set Uganda on the right path to become Africa's first oil success story.

If states have abundant natural resources yet they are unable to use them to promote growth and development, they experience what is known as the resource curse, also known as the paradox of plenty. The resource curse is paradoxical because those with limited natural resources perform, on average, better than their resource wealthy counterparts. Robinson, et al. (2006) accurately summarized the paradox:

> "Scholars of the industrial revolution and economic historians traditionally emphasized the great benefits which natural resources endowed on a nation. Paradoxically however, it is now almost conventional wisdom that resources are a 'curse' for currently developing countries".

The resource curse is not specific to any one type of resource, or to any particular region. Some resource abundant nations, such as Norway, Canada and Australia, have been able to avoid the resource curse, while others, such as Angola, Nigeria, Sudan and Saudi Arabia, have been less fortunate. Some nations have experienced declining growth and increasing income inequality due to the curse. Others such as Angola and Sudan, have experienced more dramatic effects, notably lengthy civil conflicts and extreme poverty. With the exception of Botswana, highly resource endowed sub-Saharan African nations have largely been unable to prevent the resource curse. There are many factors that cause the curse and many scholars including Ross (2002), Collier and Hoeffler (2004), Sachs & Warner (1995) and Auty (1997) among others, have written a lot about this. The factors include resource reliance, Dutch disease, crowding out effects, revenue volatility and corruption to name a few. Indeed, common-sense might suggest that access to resource wealth should be conducive to national economic development however there is a huge literature that suggests that massive resource endowments, particularly oil simply enrich a minority. The fact that oil is one of the most volatile commodities on the international market makes its marketability very unstable which may affect a country which is heavily reliant on oil unless it institutes what Stevens (2013) calls "prudent fiscal policies".

With its recent, significant oil discovery in the Albertine Rift Valley, Uganda is poised to become East-Africa's largest oil producer. Although Uganda has yet to produce oil, the commodity is already generating cause for concern. Ugandans do not have to look far to find nations plagued by the lucrative commodity. However, as will be argued throughout this paper, simply because Uganda has discovered oil does not mean it will follow the path of its oil wealthy neighbours.

In order to understand how to avoid the resource curse, it's imperative that we briefly examine the conflict, political and economic dimensions of the recourse curse and the role of institutions. We look more specifically at three economic causes of the resource curse namely: Revenue volatility; Dutch Disease; and Crowding out effects as these help us comprehend why exploitation of natural resources is often referred to as a curse.

a. The Conflict Dimension of the Resource Curse

The supposed link between natural resources and the occurrence of conflict has been widely publicised by Paul Collier and Anke Hoeffler (2005). They argue that countries dependent on primary commodities suffer a higher risk of experiencing political violence in part because the availability of resources rent would increase the financial feasibility of rebellion as is seen in Democratic Republic of Congo (DRC) and Sierra Leone (Collier and Hoeffler 1998, Collier 2007). The presence of resource rents could also increase the value of controlling the state, hence producing civil wars. The level of dependence, conflictuality and lootability of a resource also affects the vulnerability of a country to the risk of conflict (Philippe Martin; 2008, Le Billion 2001). Furthermore, it has been argued that natural resources could lead to conflicts via grievances such as inequality, forced migration (associated with resource exploitation) and unfair distribution (Humphreys, 2005).

It is important to stress that several aspects of the relationship between natural resources and conflict remain poorly understood and controversial (Ross, 2004) and the causality itself of the link between natural resources and conflict has been recently challenged. Authors such as Brunnschweiler and Bulte (2008) found that civil war creates dependence on primary sector exports but that the reverse is not true and that resource abundance is associated with a reduced probability of the onset of war. Moreover the robustness of Colliers initial results has also been challenged (Fearon 2005; Humphreys: 2005) and according to Ross (2004), the empirical evidence linking resource curse and conflict is not convincing. Nonetheless as Philippe Martin (2008) argues, this body of research helped generate various research agendas that examined the relationship between resources and violence and interesting policy suggestions have emerged such as the need to regulate extractive operations to improve transparency and reduce the risk to conflict (Humphreys 2005). Recognising the role of political actors, this paper now examines the political dimension of the resource curse.

b. The Political Dimension of the Resource Curse

Among the key mechanisms underlying the relationship between natural resource abundance and economic growth is the behavior of political elites. According to Le Billon (2005), the abundance of natural resources can lead to increased rates of corruption and rent seeking behaviour by the ruling elites. Philippe Martin (2008) also argues that the streams of wealth associated with resource exploitation create strong incentives for those in power to maximise their short term well being by capturing resource rents. It is these rents that the political elite use to buy votes by redistributing incomes inefficiently rather than engaging in bargaining with their population which creates tensions and eventually chaos (Mehlum, Moene and Torvik, 2006, Philippe Martin, 2008). In the same vein, Le Billion argues that instead of establishing coherent economic policies that maximise long term social welfare, resource rich governments often engage in clientist practices in order to manage political constituencies. He effectively summarises what the resource curse is:

> "In the absence of strong institutions and a diversified economy, large resource rents are likely to result in poor economic performance and governance failure that contrast against the high expectations of populations associated with resource bonanza. Resource dependence tends to lead to a particular kind of political rule, shaping powerful but often narrow coalitions that dampen political accountability. In their quest for power, rulers often capture and redistribute resource rents at the expense of statecraft and democracy, putting their discretionary power and fluctuating rents at the core of a political order resting on clintelism (potentially leading to) conflicts and violence surrounding resource exploitation *(Le Billon 2005)*

It is this kind of state behaviour that Uganda needs to avoid as it leads to slower economic growth and increases social tensions. I now turn to the economic dimension of the resource curse since natural resources may generate conflict through their impact on economic growth.

c. The Economic Dimension of the Resource Curse

The three main factors associated with the economic dimension of the resource curse that have been widely researched and written about are revenue volatility, the Dutch-disease and crowing out effects all of which are discussed in this section.

i. Revenue Volatility

Oil is one of the most volatile commodities on the international market. From 1974 to 1983, the price of oil increased from US $15.72 to US $61.5 per barrel, yet by 1999, the price of oil returned to near 1974 levels, at approximately US $18.5 per barrel (Stevens, 2013). The dramatic jump in revenue causes problems for states -- not only are they unable to plan effectively for the future, as they do not know how much revenue will come in from the sale of the commodity, but they are also unable to pursue what Stevens (2013) suggests is "prudent fiscal policy". Auty (1998) also concluded that revenue volatility is perhaps the most likely explanation of the resource curse for these very reasons. Ross (1999) suggests that the instability of the commodity market limits growth in resource based nations. Revenue volatility however, is only a real concern if the state exports a significant portion of its oil reserves. It is estimated that Uganda will produce 150,000-200,000 barrels per day once production reaches maximum capacity, which is quite significant.

Revenue volatility is one economic explanation why resource dependent states suffer from the resource curse. By utilizing most of its reserves to satisfy the domestic market however, Uganda can learn from the mistakes made by other African oil states and limit the effects of international price volatility. Based on current policy it appears as though Uganda is taking measures to ensure the effects of international price volatility are minimal. President Museveni has continually stated that Uganda's reserves will satisfy the domestic market before being exported[1] and Uganda continues to seek partners to provide the financing to construct an oil refinery.[2] This should arguably shield Uganda from the problem of volatility as not much of the oil is planned to be exported as crude oil. Revenue volatility is a serious concern for oil states; however, given the statements made by President Museveni, as well as the policies currently in place of building an oil refinery and satisfying the domestic market before export, it appears as though Uganda is attempting to limit the effects of the highly volatile market.

ii. The Dutch Disease

Since the decline of the Dutch manufacturing sector following the discovery of the Groningen gas fields, the Dutch Disease is used to explain why oil exporters experience stagnant or even negative growth. The Dutch Disease, according to Sachs and Warner (1995) is the most telling factor in resource dependent economies. Sachs and Warner indicate that the Dutch Disease has two primary effects on a nation's economy. First, during a resource boom (considered an improvement of the terms-of-trade or a new discovery, (Martin 2003)) the nation's exchange rate increases dramatically, and in turn reduces the nations competitiveness of non-commodity sectors, notably manufacturing and agriculture. Second, during a resource boom investment is funnelled away from those industries, because lower prices means lower profits, and is instead used to further develop the commodity market. According to Sarraf and Jiwanji (2001) by pulling capital and labour away from manufacturing, the nation experiences reduced total factor productivity, because commodities require less innovation

---

1. Biryaberma, E. & Malone, B. (2010). Uganda to start building oil refinery in 2012. Reuters News Agency, retrieved from http://www.reuters.com/article/2010/11/30/uganda-refinery-idUSLDE6AT0O620101130

2. Government plans to develop a 180,000 b/d refinery on a 29 sq. km stretch in Kabaale Parish, Buseruka sub-county, Hoima District to supply the national and regional petroleum product demand.

and labour. Although there are some case studies, such as Algeria, conducted by Conway and Gelb (1988) that suggest the Dutch disease does not always take hold, as Algeria's manufacturing sector increased, and its exchange rate declined, this simply is not the case for most major oil producing nations, as noted by Stevens (2003). In order to combat the effects of the Dutch Disease, Uganda needs to prevent currency over valuation and should utilize the revenues generated from oil to invest in other sectors, notably agriculture and manufacturing. In doing so, Uganda can avoid the negative repercussions of the Dutch disease and can use oil revenue to promote widespread growth.

Crowding Out Effects

When the state enjoys a resource boom, other sectors, manufacturing and agriculture, often find it difficult to acquire needed resources in order to develop. This is because the states, as well as business, are more likely to re-invest in the commodity already generating large rents. Le Billon (2005) suggests that developing nations mismanage resource revenues. He notes that they often fail to diversify their investment away from commodities towards other industries that could limit the effects of revenue volatility and/or the Dutch Disease. Auty (1997, 651) notes that one of the reasons why resource abundant nations economically underperform is because resources are diverted away from sustainable, growth promoting industries. Thus, although crowding out occurs in many oil-producing nations, it is not directly associated with the resource, rather is reflective of bad policy decisions made by oil states.

One can also explain crowding out effects in an international context. Foreign investors are usually not interested in the long-term development of low-medium yielding industries, rather are interested in high rents. As such, resource-abundant nations experience foreign direct investment in exploiting those resources, rather than in sectors that are more likely to generate long-term economic development. Revenue volatility, the Dutch disease and crowding out effects, are all interrelated. One cannot explain low growth rates in resource-dependent countries simply based on one of these factors, they must all be considered when examining case studies. What we do know however, is that when compared with their non-resource dependent counterparts, nations that are dependent on natural resources, have tended to experience slower overall economic growth. Uganda needs to recognize the failures made by other oil states, and needs to establish and implement policies able to prevent these economic problems. While there is variation in the literature, we can ultimately conclude that, on balance, the economic impacts of resource dependence are negative. As institutions shape the context in which political behavior takes place, the next subsection tries to examine a bit more deeply, the links between institutions and the resource curse.

iv. Institutions & the Resource Curse

According to Mehlum et al (2006) the main reason for diverging growth experiences of resource rich countries lies in differences in the quality of institutions. Institutions have been defined as the humanly devised constraints that structure political, economic and social interaction such as constitutions, laws and property rights (North 1991). Mehlum et al (2006) argue that the quality of institutions explains whether a country avoids a resource curse or not. Robinson et al (2006) also finds that the overall impact of resource abundance depends on institutions. Low quality institutions may be conducive to bad policy choices since they provide an environment that allows inefficient, politically motivated redistribution policies to take place. High quality institutions on the other hand, constrain decision makers and render rent seeking or clientelist policies infeasible or costly (Robinson et al, 2006). Esterly and Levine (2003) also argue that the quality of a country's institutions determines its level of income per capita. They utilise an institutions index taking into account: 1) voice and accountability; 2) political stability and absence of violence; 3) government effectiveness; 4) regulatory burden 5) rule of law; and 6) freedom from graft. However, while acknowledging the role of institutions and their influence on economic growth, Glaeser et al (2004) argue that they rather have only a second order effect on economic outcomes and that the first order effect derives from human and social capital. They further argue that these factors shape both institutional and productive capacities of a country (Glaser et al, 2004). In other-words institutions do matter but arguably, the policies pursued by decision makers are the major determinants of economic growth. Evidently, several scholars note that it is important to recognise the important role that institutions play in providing the right environment in which political behaviour takes place and economic policies are chosen. Uganda's politicians and policy makers need to pay more attention on strengthening rather than hampering the country's institutions and be very mindful of the kind of policies the country pursues especially with regard to managing the emerging oil industry. The decisions made by the top officials in government today will undoubtedly determine how the country performs economically.

Now I turn to examine how this literature review may provide insights into the challenges that may be involved in oil production in Uganda by first looking at its economy and then provide a brief on the emergence of the institutional framework to manage the industry.

According to the World Bank Country Profile (2010) Uganda's economy has grown at a rate of over 5 percent/year since the mid-1980s.[3] Uganda has also reduced poverty rates from 57 percent (1992/93) to 31 percent (2005/06) in 13 years.[4] On average, national incomes are rising due to the emergence of a middle class. Income inequality is still a significant hurdle for the nation, but due to sustained growth, the World Bank expects that income inequality will continue to steadily decline. The Bank, although optimistic, remains cautious about Uganda's economy for several reasons, one of them being, "...the forthcoming introduction of Oil to the nation's economy".[5]

Since coming to office, President Museveni has been able to turn one of Africa's post-independent catastrophes into what Kangave (2005) considers "...*a model for African countries.*"[6] In 1995, Uganda adopted its national constitution; in 2001, 2006 and most recently in February 2011, Uganda held national democratic elections. In 2002, the IMF's Deputy Managing Director applauded Ugandan authorities' *"ability to implement sound economic policy and structural reform, both of which have greatly contributed to Uganda's high, sustained growth rates".*[7] As well, the nearly two decade long insurgency led by the Lord Resistance Army (LRA) in Northern Uganda was ended, with the military forcing the LRA to flee to neighboring countries.

President Museveni has undoubtedly ushered in dramatic changes for Uganda, despite the many challenges still. Progress has occurred in almost all fronts; the nation is largely peaceful, the economy continues to grow, and social welfare, such as access to education and healthcare, continues to improve. Uganda's notable success in combating the HIV/AIDS epidemic is further evidence of structural change[8], although recent reports of increasing HIV/AIDS incidence rates are cause for worry[9]. Uganda is arguably currently on track to meet many of its national development goals by 2015, and is currently working towards im-

---

3. World Bank. (2010). Uganda: Maintaining Growth – Moving Towards Structural Transformation. The World Bank, retrieved from http://siteresources.worldbank.org/IDA/Resources/73153-1285271432420/IDA_AT_WORK_Uganda_2010.pdf
4. Ibid. p.1 17
5. Ibid. p.2–3
6. Op. Cit. Kangave (2006), p.147
7. Ibid. p.147–148
8. Hogle, J. (2002). *What happened in Uganda? Declining HIV Prevalence, Behaviour Change, and the National Response.* U.S. Agency for International Development, Washington, D.C.
9. A new American-financed survey held in 2012 reported that Uganda is one of only two African countries, along with Chad, where AIDS rates are on the rise. The reversal is particularly disappointing to health experts given the time and attention that have been focused on AIDS here, and the billions of dollars spent. For details see; http://www.nytimes.com/2012/08/03/world/africa/in-uganda-an-aids-success-story-comes-undone.html?_r=0

plementing a National Development Plan (NDP), building off from the earlier Poverty Eradication Action Plan (PEAP). However, the introduction of oil to the economy potentially threatens much of the progress made over the past two decades as evidenced by the resource curse literature reviewed in the first section. This is why institutions such as the World Bank, International Alert, and others, as well as Ugandan nationals (as will be noted in section four below, as well as in Dispatch, 2011) remain cautious.

There are four main reasons why there is a lot of apprehension about Uganda's emerging oil industry. First, no African nation to date has been able to utilize its oil wealth to promote economic development. Nations such as Sudan, Angola, Gabon, Congo-Brazzaville, and perhaps the most notorious, Nigeria, provide evidence that oil wealth is not beneficial. While President Museveni promotes the oil industry, Ugandans have neighboring Sudan to remind them that oil wealth does not necessarily translate into development for the poor. Southern Sudan, one of the poorest and most underdeveloped regions in the world, is also the location of Sudan's vast oil reserves which for decades have provided huge rents to the Khartoum government, yet very little for the people of the South.

Second, approximately 80% of the employed Ugandans work in the agricultural sector. Due to its high altitudes and relatively consistent rainfall, Uganda has one of the highest yielding agricultural lands in Eastern and Central Africa. Yet as discussed earlier, resource-dependent economies, or those that are transitioning to resource-dependence, often experience stagnation or even decline in their manufacturing and agriculture sectors. McSherry (2006), for example, notes that Gabon and Equatorial Guinea have enjoyed oil-related growth, yet their agricultural industries have *"...crumbled while inequality and poverty persist."* If much of Uganda's growth in the future is going to come from the oil industry, history suggests, that the agricultural sector may suffer. Overall growth may increase, as has in Equatorial Guinea, yet growth in agriculture will likely stagnate, and as such, the real incomes of the public will stagnate as well. This should be a real concern for Uganda due to the sheer number of those who rely on the agricultural sector. However, as noted earlier, Uganda, under the direction of President Museveni, has relatively improved the quality of its institutions although with many challenges still, and the living standards of Ugandans have generally improved. Thus, the problems that have occurred elsewhere may not necessarily occur in Uganda, though they should be recognized and addressed in moving forwards.[10]

Third, inflation during the 1980s reached 250 percent, yet under the current regime, it had been dramatically reduced, to 2.3 percent by the late 1990s

---

10. This is well documented in Sachs and Warner 1995.

(Musunguzi and Smith 2000). Uganda has been successful in reducing extreme inflation, yet there is evidence that the introduction of oil to an economy can have inflationary consequences. McSherry (2006) notes how inflation was one of the leading factors contributing to Dutch disease when the term was first established. In the 1970s the Dutch began commercially producing gas found near the northern city of Groningen. The "disease" kicked in when investment was drawn away from productive sectors, notably manufacturing, and was re-directed towards gas production. Inflation increased as spending increased, and the national currency, the gilder, became over valued in price. One of the reasons why real exchange rates appreciate is because of a combination of increased government expenditure and increased foreign direct investment in the resource sector. Examining six oil producers, Gelb (1988) concluded that all of them experienced increased inflation, which in turn contributed to the onset of Dutch disease. He specifically noted how Nigeria experienced extreme Dutch disease, and dramatic inflation, during the 1980s. As inflation increased, the agricultural sector stagnated, and increased general poverty. People who relied on agriculture and subsistence farming for their income lost purchasing power due to the onset of high inflation. If Uganda experiences high inflation and an over valuation of the shilling, than it will be the majority poor who suffer. Many African oil nations experience high inflation during resource booms. Yet, Uganda's experience in coping with high inflation, combined with the lessons learned from other oil states, presents an opportunity for the state to develop oil without causing hyperinflation or currency over valuation.

Over the past two decades, Uganda has sustained high growth. Based on what has been written about oil and the resource curse (see first section), notably that it often leads to high-inflation, and over-valued exchange rates, oil states often experience slow growth, increased unemployment and economic inequality – all causes for concern. Yet, as was noted throughout this section, Uganda, under President Museveni, has been able to turn a catastrophic situation into what has been arguably termed *"a model for African nations"*. Much of Uganda's success since 1986 can be directly linked to the government's ability to address severe economic problems, notably high inflation and chronic poverty. The state's ability to reduce poverty by nearly 3% per year since the early 1990s demonstrates a high level of capacity not witnessed in other African oil states. Thus, while oil can cause economic catastrophe, as has been the case on many occasions throughout Africa, we cannot assume that because oil has caused problems elsewhere that it will in Uganda as well.

Since the discovery of commercially viable quantities of oil in the Albertine Graben in 2006, the Government of Uganda has been engaged in a process to establish a robust governance regime for oil exploration, exploitation and development. Hitherto, oil exploration and development activities are governed under the Petroleum Exploitation and Production Act Cap 150 (1985) while a number of other pieces of legislation government deal with other aspects such as land and environment. However as early as 2008, Government took cognizance of the fact that there was need to develop a more robust legal and institutional governance regime to govern oil exploration and development activities for the benefit of the present and future generations in Uganda.

The promulgation of the National Oil and Gas policy in 2008 heralded the dawn of a new dispensation for oil governance in Uganda with the government committing itself to enact a new set of laws and put in place appropriate institutions to ensure good governance of the sector. In this regard, the stated policy goal of government is *"to use the country's oil and gas resources to contribute to early achievement of poverty eradication and creating lasting value to society"*. Among other things government articulated a series of principles that underpin the policy; sustainable exploitation of oil resources as a finite resource; optimising returns through efficient and effective management, transparency and accountability; competitiveness and productivity and ensuring the protection of environment and biodiversity.

The adoption of the National Oil and Gas policy was a major milestone in the Government's effort to establish a new governance regime for the oil sub-sector. However, while exploration and development activities continued and government remained committed to enter into new production sharing agreements, the legal and institutional reform process remain in abeyance. In November 2012, parliament, through a special session, stepped forward and adopted a series of resolutions demanding that government expeditiously introduces the appropriate laws necessary for the effective governance of the oil subsector. Consequently, a set of three bills were published and tabled before parliament between February and May 2012 to be debated and scrutinized as is the procedure. These include:

- The Petroleum (Exploration, Development & Production) Bill 2012[11]
- The Petroleum (Refining, Gas Processing and Conversion, Transportation and Storage) Bill, 2012[12] and
- The Public Finance Bill, 2012[13]

After a heated debate in parliament the first bill was passed in December 2012 while the second bill was passed in February 2013. The third bill is yet to be passed by parliament and the next section tries to access what the people's perceptions have been regarding Uganda's oil industry.

---

11. Uganda Gazette No. 7 Volume CV, February 2012
12. Uganda Gazette No. 8 Volume CV, February 2012
13. Ibid

4. What is in the News & what are the Perceptions of Ugandans regarding the Emerging Oil Industry?

"The discovery of commercially viable hydrocarbons in the Albertine Rift Valley in western Uganda has elicited a mixture of excitement and trepidation. For many Ugandans, there is hope that the discovery of oil and gas will result in economic transformation, growth, development and prosperity. However, for some others, there is fear, anxiety and concern that the emerging oil and gas industry presents significant challenges that the country's governance systems are not in position to effectively handle; consequently, leading to economic deterioration, insecurity and abject poverty".[14]

Uganda's Parliament passed the first of three bills relevant to the oil industry in December 2012. The bills, tabled earlier in the year, have moved slowly through parliament even though the National Resistance Movement (NRM) government maintains a strong hold over the legislative assembly. The bill, entitled The Petroleum (Exploration, Development and Production) Bill 2012, passed by a margin of 149–39. While the vote was not close, much of the frustration and negative perceptions from mostly the opposition stems from the perceived corruption surrounding the Government. A reported 5 NRM members (MPs of the current NRM government) voted against the bill, and approximately 100 MPs skipped the vote, further evidence that even members of the governing NRM regime were hesitant to support seemingly authoritative powers. Throughout this section I try to explore oil related reports from both domestic and selected international news outlets to help determine what the general perceptions are surrounding the emerging oil industry in Uganda. For the most part, I argue that oil is being met with significant skepticism and will exemplify this by focusing on four core issues; corruption, transparency, institutions and neighbouring African oil states. Ultimately I conclude that skepticism is high due to increasing corruption, weak institutions, lack of transparency and the failure of other African states to benefit from oil.

a. Corruption

Corruption in Uganda is a severe problem. Many of the articles assessed discuss particular instances of corruption, focusing on accusations against particular government ministers or discuss why corruption is likely to increase as the oil industry expands. Uganda's current Prime Minister Amama Mbabazi and Former Energy Minister Hilary Onek were accused of accepting financial bribes from firms trying to establish a foothold in Uganda's oil Industry. Both Ministers

---

14. The Daily Monitor; Oct 25th 2010; Bazira H; 'Uganda's Next Test is Oil Governance'

have also been accused of corruption in other cases, notably the Commonwealth Head of Government Meetings (2007) scandal[15]. Transparency International in its widely cited Corruption Perception Index Global 2012 report, ranked Uganda as one of the most corrupt nations in East-Africa, with a score of 2.4 on a scale running from zero to ten, higher numbers indicating cleaner government. Recent trends suggest corruption is rising rather than falling, and events such as the 2010 CHOGM spending scandal, demonstrate that Ugandans have little faith in their political leaders. NRM mainstays such as Hon. Amana Mbabazi (Uganda's current Prime-Minister) and Hon. Sam Kutesa (Uganda's current Foreign Affairs Minister) are regularly accused of corruption, and the many cases handled by the interim IGG in 2012 further demonstrate widespread corruption. The most recent scandals of misappropriation of donor funds in the office of the Prime Minister also attest to the gravity of the problem[16]. Corruption is a significant problem because it hinders, rather than helps growth, and has been associated with the onset of the resource curse in other states. According to Oredein (2012), widespread corruption is one of the key reasons why Nigerians believe oil has been a curse rather than a blessing.

Since the 1990s evidence that corruption is a serious social ill has been mounting. In order to find out more about what can explain the negative perception about Uganda's emerging oil industry; I augmented my research by conducting in-depth interviews with 34 Ugandans ranging from high level public officials, researchers, to local NGO respondents and journalists. The interviewees were purposefully selected as informants who either share a professional commitment to work on anti-corruption issues or are key players and knowledgeable about the oil industry. In order not to steer the respondents in any specific direction, I asked the respondents to describe what they think about Uganda's emerging oil industry without reservation or guidance[17].

The responses bear witness of the fact that many citizens feel the discovery of oil is going to make the government even more corrupt. As argued by the respondents, corruption in Uganda is reaching unbearable levels and oil is going to make the situation worse.

> "The government has failed to curb corruption...and now with oil money, once the country begins exporting oil, Uganda is going to become like Congo or Angola"

---

15. Mugerwa Y. (Dec 10th 2010) Uganda: Dutch government cuts UGX 10 billion aid over CHOGM Scandal. All Africa New Agency. Retrieved from http://allafrica.com/stories/201012100101.html
16. For details about this see: http://allafrica.com/stories/201301280634.html
17. The Informants were promised anonymity

"President Museveni who presides over corruption at the centre has lost the moral ground to enforce discipline among his ministers and the general public and now the oil is going to worsen the situation. In fact, as long as a corrupt civil servant remains loyal to the president and the ruling party, there is very little chance of concrete action ever being taken against them and all the money from our oil is likely to be taken"

"Corruption has become so endemic in Uganda. It is an accepted way of life. Even when someone is appointed or elected to a public office they think it is now their turn to take advantage".

"The lack of civic competence to hold our leaders accountable makes the situation worse. People especially from rural areas treat the provision of services as a gift or favour from the government. Whatever they receive from government even when they have a right to receive it, see it as a favour and are therefore all appreciative". (Interviews 2012)

According to some respondents, the emerging oil industry will make matters worse especially for the average Ugandan peasants who do not know how to demand for their rights.

"They do not see it as their right to demand for accountability when their local health centre goes without drugs for months and instead settle for anything. They never care to ask, for example, how much money has been passed to their local leaders for a road or a school. Even if shoddy work is done they remain thankful because they never expected it in the first place. Public servants have got away with a lot of stealing because they face no sanctions from the beneficiaries. In fact, corrupt people are glorified in the villages because they are the ones with money and have the capacity to solve some of their problems, such as driving a critically ill person to hospital, driving couples around on their wedding day and sometimes building churches and mosques. Some even think Oil money is going to make these civil servants richer". (Interviews; 2012).

In sum, the interviews provide an explanation as to why some perceptions among Ugandans seem to be negative about the emerging oil industry. This can at least in part be understood as a result of the high levels of corruption witnessed by the population. Thus even though the discovery of oil provides opportunities for the country, even as suggested by public choice theory, some people in Uganda agree that the costs of acting on the corrupt opportunities once they occur is likely to be very low especially when the oil money starts rolling in which makes it difficult for some, to predict a rosy future. As one respondent put it;

"People see their relatives and friends in high offices and they don't care how they get money as long as they continue to receive hand-outs....The government is very complacent so everyone is corrupt. Whether it is bad or good, the entire system in our country is corrupt. With the coming of oil, I don't expect anything to change. In fact the situation may become worse because there will be even more money to steal". (Interviews; 2012).

When reviewing the mainstream media's assessment of the oil industry it does not take one long to discover that much of the public's fears are related to systematic corruption. As oil revenues generate substantial capital for the state, the public believes that government officials and those closely aligned with the state will become better equipped to siphon off resources supposedly destined for public investment. As one Local newspaper reported, bribery in Uganda is on the increase and several Ugandans are worried about the future of the country.

"Uganda has once again emerged the country with the highest levels of bribery in East Africa, according to the East African Bribery Index 2012….the survey report…showed the Uganda Police on top of the list of bribery-prone institutions. The judiciary and land services follow in that order. Carried out in the five east African countries of Burundi, Rwanda, Kenya, Tanzania and Uganda, the survey revealed that, at 40.7%, Uganda has the highest bribery levels in the region, followed by Tanzania (39.1%), Kenya (29.5%), Burundi (18.8%) and Rwanda (2.5%). The findings are particularly disturbing because they show that the situation in Uganda is not improving. Last year, Uganda polled at 38%. What is more, the respondents (1449) drawn from central, eastern, northern and western Uganda strongly believe that bribery levels will increase in the coming years. What the results from Uganda mean is that bribery will gravely add to the cost of doing business, which in turn affects production. At a regional level, bribery will adversely affect trade between nations with countries. There is, therefore, need to address the issue. On the current state of corruption, 82% of respondents observed that corruption levels either remained as bad or increased in the last one year. The biggest reasons given for this trend were the lack of political will to fight the vice and the fact that government officials in Uganda are too corrupt to effectively fight corruption. The trend, Patrick Kayemba of Transparency International Uganda Chapter said, is worrying. "We are worse off than we were one year ago in spite of having the best anti-corruption institutions in the region. This trend is very, very worrying (The Observer Newspaper, 2012)

One of the most contentious issues discussed widely in the media late last year in 2012 was one of the legislative bills governing the oil industry. This bill saw the granting of *"authoritarian-like"* powers to the Minister responsible for oil, otherwise known as Clause 9. While the NRM originally removed some of the authority given to the minister in November, this position was reversed and the legislation was passed in December of 2012 as it was originally drafted. According to Clause 9, the Minister will now be able to approve or reject licenses for oil exploration and drilling as he/she sees fit. The passing of the clause asserts that the most important body is the Minister for Oil and not the legislative nor the executive branch of government (although the President, and likely cabinet, will have to support new deals). The Petroleum authority's independence is compromised by a requirement to adhere to any directions by the minister. The fear amongst many Ugandans is fairly straightforward; as the Ministers authority over the industry increases so too does the potential for the abuse of power. The Minister evidently wields too much power and has the leeway to do just about anything. Without stringent checks and balances, the Minister will have the opportunity to take advantage of his/her position in government, and in doing so may personally squander national resources.

However, fears over corruption are not limited to the Minister for Energy / Oil. There are also widespread fears amongst the public that due to the apparent rise in systemic corruption, the government and those it colludes with, may squander the forthcoming resource windfall. Many western nations, notably the UK, Ireland and Norway have recently withheld substantial foreign aid as it is believed that the government has embezzled funds originally earmarked for post-war reconstruction in Northern Uganda. According to The Guardian in a November 16, 2012 publication, the U.K. was withholding an additional £11.1m of aid due to fears over financial abuse. The African Review Newspaper of 4th Dec 2012 also reported that the EU, the United Kingdom, the World Bank, Austria and other countries had suspended up to $300 million promised in budgetary support each year, up to 2013. In October, Sweden and Ireland had also suspended project support aid to Uganda over alleged corruption in the Office of the Prime Minister (OPM) and sent a team to investigate the matter. The move followed a draft report by Auditor-General John Muwanga, which found that at least Shs50 billion in aid from Ireland, Norway, Sweden and Denmark had been misused[18]. The positive feature about all this corruption scandal in government is that it was discovered by the Auditor General's office which suggests that perhaps the institutions in Uganda if left to operate, would enable the country avoid the resource curse. Often, it is the donors who raise the flag but in this case, the financial mis-use was discovered by the Auditor General's office which is a positive development. Arguably, such an initiative by the Auditor General's office in Uganda, could never have happened ten years ago.

The government of Uganda, once heralded internationally for its ability to combat the HIV/AIDS crisis and President Museveni's original commitments to a free and fair electoral process, has been on the receiving end of substantial foreign criticism. The Uganda government will have to tighten its financial belt significantly with this withdrawal of donor funding. According to the OECD Uganda received $1.73billion in foreign aid and international development as-

---

18. For more details see: http://www.africareview.com/News/Donors-cut-all-direct-aid-to-Uganda/-/979180/1636076/-/11k5ymv/-/index.html

sistance in the year 2010.[19] The amount of aid donated to the country represents a very substantial amount of the country's annual GDP. As concerns over corruption rise internationally, the richer nations may be less willing to contribute important foreign aid.

Corruption in Uganda is undoubtedly a widespread, national problem. In a 2012 report Transparency International noted that Uganda has become the most corrupt country in East-Africa (a group that includes Kenya, Tanzania, Rwanda and Burundi). As noted above, corruption is one of the most consistent themes found in both of Uganda's leading dailies, the Daily Monitor and New Vision, and is a consistent topic in some of Uganda's widespread weekly publications, including The Independent. Fears over corruption permeate almost all aspects of Ugandan society, and it is clear when assessing both domestic and international media, that rising corruption in conjunction with increasing oil production is a significant cause for concern. The highly publicised debate over Clause 9 demonstrates Uganda's general fears that oil will only further solidify state sponsored corruption rather than promote national development[20].

Since discovering major oil deposits in 2008 the Ugandan government has consistently argued that the resource will help propel Uganda into a new era of development. However, noting increasing corruption, rising tension between regional authorities (notably the kingdoms), as well as the general secrecy surrounding the industry, oil remains a politically turbulent issue. A simple review of Uganda's two leading daily journals, the Daily Monitor and New Vision since 2010[21], demonstrate that mainstream skepticism exists. The following few quotes further demonstrate some of the many reasons why Ugandans appear to be skeptical of oil.

On November 11th 2010, in an Article published by the Daily Monitor Newspaper entitled: First Family too close to Oil Sector, Mugerwa noted:

> "From a governance perspective, the military control of the oil exploration areas by two of Museveni's close relations is evidence of the increasing personalisation of control by Museveni of the oil and gas sectors. Such deviation from democratic principles at this stage is highly undesirable".

In another article published on November 22nd 2010 entitled: "Slippery Politics of Uganda's Oil" Angelo Izama stated that:

---

19. For the full story see: http://www.oecd.org/dac/aidstatistics/1883200.gif
20. For details about the heated debate see http://allafrica.com/view/group/main/main/id/00020853.html
21. 2010 was chosen as a benchmark year because official elections for the 2011 elections in Uganda were launched that year which generated a lot of debates in the media relating to the oil industry up to today.

"Uganda's oil is bleak. It is unlikely that oil revenues will be better managed and may be that is where the focus of opponents of the incumbent president ought to be laying emphasis while campaigning on this subject".

In *"Where are Ugandans in Lucrative Oil Deals?"* (2011), Chris Obore notes how many local firms, capable of providing services for the infant industry are losing contracts to foreign companies. The Minister responsible was quoted as saying that Ugandans have been "sleepy to date" while foreign firms have mobilised. Critics however have claimed that foreign companies have benefitted from insider information within government through corrupt officials and that many local companies have been left in the dark[22].

Overall, oil related articles published in Uganda's media since 2010 paint a rather pessimistic picture of the future of this emerging industry, Much of the coverage focuses on how Uganda could become the next Nigeria and how Ugandans are unlikely to prosper from the industry and how oil could return Uganda to its once violent past.

It should be noted that a recurring theme in much of the press coverage of the oil industry in Uganda even for the most optimistic articles, is that the state needs to be more transparent, more accountable and needs to systematically combat corruption. If the curse is to be avoided, "oil deals must be open and transparent.."[23]

Corruption is undoubtedly a significant cause for concern because corrupt officials misuse resource windfalls for personal gain[24]. Oil corrupts and oil states usually have higher overall rates of corruption than non-oil states[25]. As revenue is mismanaged and resources are not used to promote pro-poor growth, citizens will become increasingly disgruntled with the state, especially those living in regions disconnected to the national government. Endemic corruption in Uganda has the capacity to derail the country's future growth prospects. The government would do well to address and combat widespread corruption in order to appease the public and to put oil profits to good use.

---

22. Obore Chris (2010) Where are Ugandans in Lucrative Oil Deals? The Daily Monitor: Retrieved from http://www.monitor.co.ug/News/National/-/688334/1074308/-/cjyb6Iz/-/index.html
23. Kasita 1 (2001, Feb. 2010) Oil deals should be open says Law Journal. New Vision Retrieved from http://www.newvision.co.ug/D/8/220/746270/Oil
24. For more recent discussions on Corruption in Uganda, see http://www.guardian.co.uk/katine/2009/mar/13/corruption-endemic-in-uganda
25. Ross (2002) further solidifies this argument by presenting the case of Angola, one of Africa's largest Oil producers. Ross notes how in 2001 Angola lost approximately $ 1 billion USD due to corruption, a staggering amount considering it is one of the least developed countries in Sub-Saharan Africa.

b. Weak Institutions

Another important sentiment that emerges when reviewing mainstream media and having interviewed several Ugandans is the lack of trust in institutions. While this could be considered corruption, here I consider a lack of trust in institutions as a belief that they are inefficient and unable to govern such a complicated resource as oil. Although Uganda has seen marked improvement in a number of sectors, and continues to invest in strengthening public institutions, there is still significant progress to be made. Ugandans seemingly believe that institutions established are not effective enough to properly govern, nor have the capacity to deal with systematic problems, including corruption. As one respondent from civil society noted;

> The government has decided to set up a petroleum authority for the governance of oil as if its an ordinary authority yet such an institution is very strategic as it is supposed to oversee a finite resource. If the government is setting up the same kind of authority like the national forest authority or the wildlife authority or our scandal ridden investment authority, then our country is doomed. At the moment most of these authorities are facing one crisis or another. (Interviews, 2012)

Although there has been a recent improvement in service delivery, notably electricity, Uganda still lags behind when providing basic services, such as access to basic healthcare, proper infrastructure and road development. Uganda's public education system, though having improved over the past decade, remains well behind those of its regional neighbours. In Uganda, those with money prefer to send their children to private educational facilities because it is generally believed that publicly run institutions provide a worse education. The general concern is that if Uganda is not able to provide a decent public education system, something it has been responsible for since 1962 than how can it properly govern and manage an emerging oil industry. As one respondent put it;

> "We have failed to manage most of our key sectors which are so fundamental for our transformation like Education, and we don't even have a plan on how we are going to create the requisite human resource to manage this industry, so how can you expect us to manage this oil? All our institutions are weak"(Interviews, 2012)

Oil states have significantly weaker institutions than non-oil states, Terry Karl (1997) suggests that oil states have little interest in developing sound institutions because stronger institutions demand higher accountability. Acemoglu, et al (2003) also argue that weak institutions are chiefly responsible for the lack of growth and development in much of the global south. Inefficient public systems, especially the ability to collect revenue (tax systems) limits the state's ability to deliver public goods it is specifically responsible for. Though Uganda went through decades of instability following independence it has been nearly 30 years since the current administration came into office. Though improvements have been made, Uganda's institutions remain quite weak, especially the electoral commission and the tax authority, two of the principle institutions of any modern democracy.

Two conclusions can be draw from the above; on the one hand, Ugandans evidently do not trust their institutions, and two, public institutions have a difficult time delivering the goods and services they are supposed to.

b. Lack of Transparency

It is rather unfortunate that the oil debate in Uganda has been shrouded in rumours and a lack of clear information. This has been the case especially with regard to the petroleum – sharing agreements signed by the government and associated allegations of bribery[26]. The government has of today, released only partial details of the Production Sharing Agreements (PSAs) to parliament[27] but has not disclosed these to the public. This lack of transparency has been divisive and arguably unnecessary. It may well be the case that these PSAs could have been well negotiated but the fact that nothing has been made public creates a lot of suspicion and negative perceptions. These have continued to be the centre of speculation which underlines the risks of controlled information too closely.

As one respondent put it;

> "The fears of civil society are genuine. We have been kept in an information-vacuum. The oil contracts are designed in secrecy. As a civil society activist, I feel cheated. I am worried that the final petroleum law may come up without adequate inputs from a wider section of Ugandans. If we were brought on board properly, there would be no suspicion. There is no transparency in the conduct of oil affairs. The oil issue should be treated as an issue for all Ugandans". (Interviews, 2012).

The government should heed to this and ensure that access to information is enshrined into law particularly for the oil industry. As Steven's (2013) has argued, in many countries around the world, PSAs are now released as a matter of course. Though full disclosure of contracts is a relatively new phenomenon, driven by campaigns such as 'EITI and Publish what you Pay', countries like Ghana and Liberia have enshrined this into law. Many more countries like

---

26. For more details see: Platform, 'Uganda's contracts – a bad deal made worse', http://www.platformlondon.org/carbonweb/documents/Ugandas_oil_contracts_A_Bad_Deal_Made_Worse_Tullow_Heritage.pdf.
27. Global Witness, 'Civil Society Groups Challenge Ugandan Government Over Oil Transparency', 10 July 2012.

Egypt, Georgia and Krygyzstan give parliament the right to approve all contracts (Stevens, 2013). In Nigera, EITI has led to audit report that have placed immensely rich data and information in the public domain thereby strongly empowering civil society to hold government to account[28]. Although Uganda has in principle committed itself to EITI membership, it has not taken any steps yet to become included. Uganda also legally recognises the rights of citizens to see information held by government as enshrined in the Access to Information Act (2005) but this has not been fully operationalised yet and is contradicted by the provisions for confidentiality in the New Petroleum bills.

Shepherd (2012) in his piece, entitled 'Oil in Uganda: International Lessons for Success', discusses at length how Uganda has the potential to succeed, and has many advantages, notably time, lessons learned from others and fairly sophisticated technical knowledge of oil, that other oil states have not had. Shepherd argues that "Uganda is in the advantageous position of being an established democracy, with enshrined legal and media freedoms."[29] Shepherd states that Uganda has been able to develop a relatively harmonious sense of community, and though he notes social and ethnic divisions remain, he argues that they do not permeate society.

Though Shepherd should be noted as one of a few authors who have taken the oil debate from a positive perspective, some conclusions he generates are disturbing. While Uganda can be considered a basic democracy as elections do happen fairly regularly, it cannot be considered a well-established democracy, similar to the likes of Chile, Norway or Botswana. Shepherd notes that one of the challenges to Uganda is the lack of strong political opposition, a key component of any well-established democracy. The level of harmony between ethnic groups in Uganda and the sense of general community sentiments expressed by Shepherd are also debatable. Uganda is a highly divided society along many lines, ethnic, social, economic, geographic, religious, to name a few. How oil revenues are spent certainly has the ability to exacerbate, rather than reduce tensions along these lines. More importantly, and as Shepherd notes, the government should attempt to limit the misperceptions about oil revenue sharing and be very transparent about the oil deals and contracts signed[30].

---

28. For more information, see EITI Factsheet No.1, http://eiti.org/files/4-jan-2013-EITI-Fact-Sheet.pdf.
29. Shepherd Ben (Feb, 2010) Oil in Uganda: International Lessons for Success, Chatham House. Retrieved from: http://www.chathamhouse.org/sites/default/files/public/Research/Africa/0113pr_ugandaoil.pdf
30. For more discussions about lack of transparency in the oil laws in Uganda see; http://www.theeastafrican.co.ke/news/Uganda-new-oil-law-is-silent-on-transparency--/-/2558/1650006/-/iv55d8z/-/index.html

d. Failure of Others

Problems such as the dutch disease, over inflation, poor terms of trade, and many other economic concerns have plagued many nations reliant on resource extraction. Indeed it is the failures of many nations to properly utilize highly sought after resources that has resulted in a litany of resource curse related literature. The inability of states such as Nigeria, Angola, the DRC, Sudan, Libya, et cetera, to use their vast resource wealth properly has cast global skepticism over oil. Uganda is arguably no exception to this rule. But as we will examine in the case of Botswana, not all states fail with abundant natural resources. Resource extraction has a clearly mixed track record. However, based on the interviews conducted and the media, it appears as though Ugandans are more skeptical, rather than optimistic when it comes to oil. While one is unable to specifically conclude how the failures of other states, especially other African oil producers, has impacted general public opinion, the fairly regular reference to failed neighbours suggests some impact.

As one of the respondents noted:

> "All our Countries in Africa have failed to manage their oil resources productively. Nigeria is a perfect example of a country that has suffered because of oil and the list goes on including Angola, DRC and Sudan. When we have selfish leaders in power all over Africa, I don't expect Uganda to become a miracle success". (Interviews; 2012)

The Uganda government has however repeatedly criticized comparisons with Nigeria, Angola and other resource-curse countries. It has raised expectations that Uganda will be Africa's new Norway. As Kizza et al (2011) have argued, the government originally used exaggeration and/or a laissez faire strategy to respond to people's expectations and this has created disillusionment. The president, for example raised people's expectations when he emphatically argued, as early as 2006, that Ugandan oil will be a blessing, not a curse:

> "There is a lot of nonsense that the oil will be a curse. No way! The oil of Uganda cannot be a curse. Oil becomes a curse when you have got useless leaders, and I can assure you that we don't approach that description even by a thousandth of a mile… The oil is a blessing for Uganda and money from it will be used for development"[31].

On 9 October 2009 (Uganda's 47th Independence Day) the president asserted that:

---

31. See IOL, South Africa, Uganda Announces Oil Discovery,
    http://www.iol.co.za/index.php?set_id=1&click_id=68&art_id=qw1160371442337B225

"No one, in Uganda or internationally, can now doubt the country's steady and deliberate path to a middle income country status in the near future,' thanks to the discovery of oil"[32].

A year earlier, the president had announced that Uganda will not export crude oil (Juuko and Odomel, 2008). It is now over four years and no oil is flowing yet. People's expectations have arguably been dampened now, more so by sceptics who have warned that oil could lead to the displacement of "private sector-led growth" with a state-driven economy. It has been argued that Uganda's fledgling democracy could descend into an overt dictatorship as predatory political elites position themselves to enrich themselves at the expense of the country as has been the case in Nigeria, Angola and Sudan (Kizza et al. 2011).

But what do the bills say and where are the gaps? I now turn to look more specifically at the new oil bills and try to provide a critique.

---

32. Uganda Media Centre, Independence Speech, 9 October 2009, http://www.mediacentre. go.ug/details.php?catId=6&item=634

5. Important/Relevant Clauses in Policy Documents relating to Uganda's Oil Industry

In this section I try to explore the three new bills pertaining to oil and oil management in Uganda and try to highlight some contradictions. Only two (The Petroleum Bill 1 & 2) of these three bills have been passed into the law. The first bill was passed with what can be considered significant public skepticism not reflected in the overwhelming legislative support. Here I attempt to assess each bill independently, discussing the particular clauses I believe to be relevant to the management of oil and oil revenues.

The first and most significant bill, entitled Bill No.1 *'The Petroleum (exploration, development and production) Bill,'* focuses specifically on the creation of the Petroleum Authority of Uganda, a National Oil Company, and the specific details surrounding exploration, development and production. There are two particularly important facets of this bill; first the creation of the Petroleum Authority, which will regulate exploration, development and production of oil, and second, Bill No. 1 will establish a National Oil Company which will "...manage Uganda's commercial aspects of petroleum activities and the participating interests of the State in the petroleum agreements." According to Clause 44, the National Oil Company will;

- handle the state's commercial interests in the petroleum sub-sector,
- manage state participation in the petroleum activities;
- manage the marketing of the country's share of petroleum received in kind;
- manage the business aspects of state participation
- develop in depth expertise in the oil and gas industry;
- optimise value to its stakeholders;
- administer contracts of joint ventures;
- participate in meetings of licensees; and
- investigate and propose new upstream, midstream and downstream ventures initially locally but later internationally

There is little further insight as to what specific functions both the Petroleum Authority as well as the National Oil Company will carry out. The bill does not present a clear separation of different roles to be performed be the different institutions. What is becoming clear however is that the government is interested in setting up a number of different agencies designed to govern different aspects of the industry. While this may seem prudent, we must remain cautious as what these agencies actually do on a day to day basis may overlap, waste resources, and could potentially destabilize the management of both the revenue and the resource. When something goes wrong, competing agencies will likely look to each other to lay blame unless the roles and responsibilities of each are clearly laid out and adhered to. It currently appears as though the Petroleum Authority will focus on regulation, while the National Oil Company will actually engage directly in the industry on behalf of the government; it remains unclear where the responsibilities of one stop and the others start. It is, of course, crucial to regulate the oil industry and it is prudent of the Government to create such a regulatory authority. In order to be a truly effective agency, it must be clearly laid out how the authority will be separated from direct government involvement. If the National Oil Company plays a role in regulation than we could arguably soon see rules and regulations bending to suit the interests of the company.

In addition, Bill 1 provides excessive powers to the Minister. The powers vested in the minister are arguably far reaching and according to some NGOs in Uganda...; "generally inconsistent with the fundamental principles of checks and balances which are the cornerstone of an effective governance regime for the oil subsector"[33]. The bill spells out the minister's powers to include issuing of licences, drafting legislation and developing regulations. This contentious bill gives the minister of energy unlimited powers to negotiate, grant and revoke oil licenses. There is also potential for confused lines of authority. Many countries that have failed to utilise oil for the benefit of their citizens have similar institutional structures where substantial decision making powers are vested in a single institution with very limited checks and balances. Uganda can do better to ensure that there are relevant checks to the powers of the minister and prevent any excesses that may lead to abuse.

Nevertheless, Bill 1 is a positive step, building significantly on the previous bills relevant to the oil industry. The oil industry needs stringent regulation and it must have a dynamic agency ready and able to implement regulations. Whether or not this agency will become dynamic and will be able to carry out its mandate in a positive fashion remains to be seen. The creation of a National Oil Company also falls in line with prudent practice and has been done by many oil producing states. The government needs an arm in the industry able to carry out its interests, as long as those interests serve the general interests of the public, not those in government. Clause 18.2 provides a positive sign, although the clause gives the minister the powers to appoint all of the Board of Directors for the regulatory authority, as it stands, all appointments must be approved by the entire cabinet.

The Uganda parliament, on 21st Feb 2013, passed the second Petroleum bill moving closer to completing the new regulatory framework for the nascent oil sector. This midstream bill, known as the Petroleum Refining, Conversion, Transmission and Midstream Storage Bill 2012, was passed nearly three months after the upstream oil bill was passed. The midstream bill will regulate midstream operations, including refining, transportation and storage of oil prod-

---

33. Interview with an NGO activist in Kampala, October 2012.

ucts, once the country starts production. Unlike the previous upstream law, whose passing was delayed for several months as parliament debated the executive over the powers of the minister to license and revoke licenses, the midstream law quickly sailed through parliament after a few weeks of debate. Activists criticized the bill, however, saying once again it gives sweeping powers to the minister, which is a threat to good governance.

A member of a pressure group; "Publish What You Want" noted;

> "It is going to be hard for Uganda to avoid the oil curse given the manner in which these bills have been passed,[34]

The bill grants the minister sole powers to award, suspend and initiate the development and implementation of policies concerning midstream operations among others.

While Petroleum Bill No. 1 focuses more specifically on the creation/roles of various government agencies, Bill No.2 focuses on the partnerships between government and licensee's, otherwise known as those granted licenses to explore, drill for, and produce oil. It is entitled Bill No.2 *'The Petroleum (refining, gas processing and conversion, transportation and storage) Bill,'*. Clause 54.1 provides some optimism that the government is listening to concerns of the public. The clause outlines how licensees and contractors must give priority to Ugandans and registered Ugandan companies for the provision of goods and services while supporting their commercial endeavours. It is prudent of the government to motivate licensees to utilize local expertise, however there are some shortfalls. First, it needs to be established what is considered 'competent' and who decides whether or not an individual or company is 'competent'. Second, what incentives are there for licensee's to utilize locals. A tax break for example would likely motivate companies to take advantage of local competence/invest in developing local competence. It is only logical that the government wants licensee's to employ locals, after all it is the locals who decide who governs (in a true democracy). Yet the government could build on this clause and provide further incentives. Clauses 54.3, 54.4, 55.1, 56.1 and 56.2 all go into further detail as to how licensees should interact with local populations and how interaction should be reported. For example, 56.1 states that "A licence shall include a clearly defined training programme for the local employees of the licensee, which may be carried out in or outside Uganda and may include scholarships and other financial support for education". This clause makes it clear that the government will require licensee's to utilize local labour and to invest in local populations, too what extent however, remains unclear.

---

34. The Wall Street Journal, Friday 22nd Feb. 2013; ' Uganda Parliament Passes Second Oil Bill' Retrieved from : http://online.wsj.com/article/BT-CO-20130222-704832.html

Clause 54.4 describes how licensees shall provide a report, to the Petroleum Authority, detailing the use of Ugandan goods and services during the calendar year. The clauses mentioned all describe how the government is looking to include Ugandans in the oil industry and develop 'local content'[35]. This is a step in the right direction, and falls in line with President Museveni's declaration that over 100,000 jobs would be created directly or indirectly because of oil. However, there are further steps that need to be taken by the government to ensure licensees involve Ugandans to the extent desired. Tax breaks and financial incentives for inclusion are a potential path the state could go down, as are financial penalties and repercussions for not involving/utilizing Ugandans. One problem that could arise is due to expectations; what the government considers appropriate local inclusion could be far off from licensee expectations. Numbers should be clarified throughout the negotiation phases, before foreign firms begin operating, so as to avoid future problems.

While the clauses discussed above present a positive image, there is still some cause for concern. Clauses 76.1 (a through d) and 77.1 describe ministerial oversight and responsibility. According to clause 76.1 "The Minister shall, subject to confidentiality of the data and commercial interests, and in accordance with the Access to information Act 2005, make available to the public –

- Details of all agreements, licenses and any amendments to the licenses or agreements whether or not terminated or valid;
- Details of exemptions from, or variations or suspensions of, the conditions of a licence;
- Licenses; and
- All assignments and other approved arrangements in respect of the licence

To date (as noted earlier) very little information regarding the details of any agreements between government and oil companies, including licenses and any other arrangements, have been disclosed. It is well known and highly documented, and was discussed at length earlier in this paper that one of the main concerns regarding oil in Uganda remains to be the general secrecy of the industry. We do not know what levels of revenue sharing schemes the government and various firms have agreed to, nor do we know much else what was signed between the government and all firms involved. The preface that data will only be shared if it is not considered confidential remains highly problematic. We do not know what constitutes confidential, nor know who decides what is and what

---

35. Local content in this case basically means that anything within the oil sector (goods/services) that can be provided by Ugandans should indeed be provided by Ugandans. Only the extremely specialised works may be carried out by foreigners. But, even for these 'specialised' services, Ugandan participation is encouraged.

isn't considered confidential, though it could be presumed to be the Minister responsible and the remaining executive.

Clause 77.1 goes even further, stating that all data and information submitted to the government from licensee's will remain confidential and will not be reproduced or disclosed to third parties. We can assume from this clause that we will not know how much oil will be extracted, and therefore will make it very difficult to determine how much revenue the government should be receiving from each firm, thus making it difficult to follow the money trail. The government needs to look to the examples set by those who have turned resource abundance into prosperity, such as Norway, Canada, and on the continent, Botswana. Initiatives like the Extractive Industries Transparency Initiative (EITI) or Publish What You Pay further demonstrate why information should be made public. In general, the more information that is made available, the more accountable the public can hold the government. If the government is serious about making oil wealth work for Ugandan's than it should start with reporting what wealth is being generated. It is unlikely that the citizenry, who in general know very little about oil and oil economics, would have too much to complain about if the information was presented to them. Some have speculated that the government does not want to make agreements public because they are not as financially lucrative as some have suggested (including the IMF). This in itself is not reason enough to keep information of this nature private. On many occasions the state has referred to oil as a public resource. If this is in fact true, then it would be advantageous, and from a long term perspective, beneficial for all Ugandans to be publicly aware about the resource. EITI was directly established to help promote transparency because, as was discussed earlier, and has been forwarded by the likes of Collier, Sachs, et cetera, transparency is key to proper resource management.

In the second half of 2012, Parliament began debating 'The Petroleum Bill 1 of 2012' and though it was passed in a December vote in Parliament, it was not done so without significant contestation, both in the political and public spheres. Most controversially of all the clauses of the Bill was the now infamous clause 9, which asserts the Ministers responsibilities for granting and revoking oil licenses. Over the years Ugandans have become increasingly skeptical of state sponsored corruption, the 2005 removal of presidential term limits as well as the 2010 CHOGM are but two examples of many hotly contested political 'scandals'. Indeed it is not uncommon to read about bureaucratic and political corruption in Uganda on a daily basis in both local and international media. Clause 9 has created increased fear over the role of government in oil. As has been discussed earlier, one of the primary concerns about the resource has been the limited amount of information shared with the public. The inclusion of 'clause 9' appears to suggest even less information will be made available, and demonstrates an increasingly involved role of the Minister vis a vis a less involved role for Parliament. While we are still at least two years from full scale commercial production, the government should attempt to dilute political hostilities with regards to oil, and re-examine clause 9. As oil becomes an increasingly important component of Uganda's society, more and more Ugandans will keep a close eye on the minister, and if clause 9 remains upheld in the forthcoming years, President Museveni would do well to appoint his most trusted colleague and one well respected by all facets of society. Appointing any of his tainted colleagues in this position would surely result in further instability down the road.

The last of the bills remaining to be discussed is the Public Finance Bill. It was introduced in early 2012 but is still to be considered. It was developed to simplify and bring together all financial related legislation into one prevailing document. One of the many significant sections of this Bill discusses oil wealth and management. Here I briefly discuss what the Bill, as it stands, will do to improve oil management. In that same light I also echo some recommendations made by others, notably Revenue Watch International, which would help Uganda adhere to known best practices.

Chapter 7 of the Public Finance Bill takes significant steps forward in addressing how oil related revenue will be utilized by the government. Notably, the Bill, if passed, will result in the creation of a National Petroleum Fund, which will be tasked to manage the wealth generated from the industry. Additionally, the Bill recognizes the importance of proper revenue management, legislating that the revenue generated shall not be used to accrue new loans or debt, as has been done on many occasions elsewhere. The Bill also establishes how and where revenue will flow, and what standards will be set up to ensure those who owe royalties to the government will actually pay them. Lastly, chapter 7 establishes basic transparency and management practices which should help the still infant industry limit problems experienced elsewhere when it comes to revenue management. In general, the government should be applauded for the steps taken in the Public Finance Bill as they are not only generally good practices but they are steps, recognized by many, that could lead to increasingly stringent rules and regulations governing the resource. However, while Uganda is largely compared to its regional neighbours when it comes to oil regulation, and thus the expectations are significantly lower, Uganda should aim to include additional legal frameworks which would make the state more comparable with nations such as Norway, Canada, Botswana, et cetera. Revenue Watch International (RWI) tabled four key suggestions which should be taken seriously by the Government of Uganda (GOU) if they wish to be seen as Africa's most forward thinking revenue manager. These recommendations are, The GoU should 1) address oil revenue volatility, 2) Ensure proceeds from oil revenues are invested for growth and development, 3) strengthen oversight and transparency rules to promote accountability and 4) establish a widely acceptable rules-based and transparent revenue scheme.[36] The pre-eminent concern with oil management in Uganda is how the wealth generated is going to be used. While land disputes and environmental concerns, as well as foreign involvement, are all paramount issues, there is no issue as important as wealth management. If oil wealth is used for prosperous means and Ugandans, on average, see improvements, then oil will be a success story. If wealth is not used appropriately, and further exacerbates existing corruption, Uganda could see the onset of the resource curse, leading to a multitude of problems discussed earlier.

Chapter 7 of the Public Finance Bill does represent an improvement of oil wealth management which should not be understated. However, as RWI discussed, further steps can be taken to improve the quality of the legislation which would likely result in better management practices.

Although attempts have been made to improve the legislation for Uganda's oil industry, there are still areas of concern as highlighted above, which government would do well to re-visit and address accordingly. In order to avoid the resource curse, administrative effectiveness, respect of rules and public trust in the government are all necessary pre-liquisites.

I now look at a country whose standard of living was below that of Uganda in the 1960s but has shown that mineral wealth can be used to sow seeds of development in other sectors. It is from such that Uganda can also draw lessons for developing the oil industry and avoiding the resource curse.

36. http://www.revenuewatch.org/sites/default/files/RWI_comments_PRM_bill_Uganda.pdf, pg 4

The Botswana Success Story; what explains this & what lessons can be learnt?

Botswana has been referred to as a successful democratic developmental state in Africa (see Leftwitch; 1995, Edge; 1998, Taylor; 2003, Sebudubu; 2005, Mbabazi & Taylor; 2005). Once a poor nation, Botswana has, over the past four decades, become one of Africa's success stories. In order to account for how Botswana turned poverty into wealth and resources into blessings, we must ask three key questions. First, what are the key factors that account for Botswana's relatively good performance? Second, with most of the other resource rich African Countries, like DRC, Angola, Nigeria , et cetera, experiencing poverty, failure of institutions and increasing corruption, what accounts for Botswana's positive trajectory? What are the characteristics of Botswana's institutions and leadership that have permitted the country to remain relatively immune to many of the predicaments affecting resource rich African countries? Lastly, what lessons can be learned from Botswana for the benefit of Uganda?

According to Sebudubudu & Molutsi (2011), the nature, character and behavior of the elite in Botswana have a lot to tell about the country's success. The unique politics and governance style of the ruling elite in Botswana has shaped coalition building and networking among the leaders of the country and led them to frame out a collective vision for national development. Several other scholars like Hillbom, 2008 and Mehlum et al, 2006 have also argued that the development of the ruling elite in Botswana has played a key role in shaping the country's development trajectory and contributed to the successful development of the state. Diamonds are undoubtedly the major reason (economically) behind Botswana's transition from one of the poorest countries in the world at independence to its current middle income status. But Botswana's experience is undoubtedly an anomaly on the continent especially when compared to countries with rich natural resources like Nigeria, Sierra Leone and the DRC, which have been plagued with weak institutions, corruption, neo-patrimonialism[37] and conflicts to mention but a few. So what is unique about the elite in Botswana that seems to have evaded other African countries?

Paul Collier has tried to demonstrate the linkage between resource rich nations and conflicts (Collier and Hoeffler, 2004) and other authors, like Bates (2008) and Ron (2005) have provided more viable arguments focusing on the role of state institutions and the distributional choices made by the rulers/authorities or the elite which arguably triggers political instability and leads to violence and state failure. Among the many key reasons given for state failure are predatory behaviors of incumbent elites who embezzle public wealth, engage in nepotism, favoritism and authoritarianism. According to Bates (2008), incum-

---

37. Neo-patrimonialism is a very extreme, exploitative, personalized and highly centralized form of exercise of state power very evident in most resource rich nations like Uganda.

bent political elite tend to turn into predators of their own people because of their perception of the risk of being overthrown. He argues and quite convincingly, that once a leader feels threatened of losing office, this creates a decisive incentive to prey while still in office. This could arguably explain the recent increase in state sponsored corruption witnessed in Uganda today[38].

Building on Bates' arguments to understand Botswana's success, the next sub-section takes a closer look at Botswana's development trajectory; the behavior of the state and its institutions as well as its actors and tries to analyze the socio-political and economic environment in which the state was built. This, according to Bates, is what significantly affects the incentives and constraints faced by many leaders to develop their countries.

a. Botswana's Political Economy

Unlike the majority of African states, Botswana has never experienced any insurgencies or civil conflicts in its post-colonial history. This African example has rendered Paul Collier's (2006) argument that *'exploiting massive amounts of mineral wealth, mechanically leads to chaos'* very debatable. The poor economic achievements of resource-rich countries like Sierra-Leone or DRC are increasingly seen as consequences of weak institutional design, lack of accountability and corruption rather than abundance of mineral resources. The country boasts a good human rights record and has been cited by many as Africa's model of good governance.

Botswana has also had unparalleled stability with its ruling party; the Botswana Democratic Party (BDP), which has enjoyed uninterrupted state power since independence. It continues to enjoy overwhelming support and although other political parties exist, they do not pose any serious challenge to the BDP. More importantly, not only has Botswana avoided political turmoil, it has also achieved outstanding economic and social progress[39]; again contradicting

---

38. Corruption in Uganda has become very perverse. Recent statistics from Transparency International indicate Uganda as the 45th most corrupt country in the world. In August 2012, the East African Bribery Index conducted by Transparency International, Kenya ranked Uganda with the highest level of bribery in East Africa at 40.7 per cent. According to the report, to access most of the essential services in Uganda, you are more likely to fork out a bribe than in any other East African country. The report comes at a time when the country is awash with revelations of abuse of office and corruption in various ministries, the police investigation department currently probing cases of alleged cases in about five ministries. Most prominent is investigations into alleged corruption in the Office of the Prime Minister, where it is believed that more than Shs50 billion meant for peace recovery programmes in northern Uganda was embezzled during the period 2010–2012.

39. When Botswana became independent in 1966, it ranked among the poorest and relied heavily on South Africa. It had only 12 kms of paved road and 22 University graduates. Its per capita GDP has however grown from 60 USD$ in 1966 to 3600 USD $ in 2002 by which time the World Bank had classified Botswana as a middle-income country (Battistelli S & Yvan Guichaoua, 2012).

Colliers argument that resource rich countries often experience low economic growth patterns.

When Botswana became independent it was one of poorest countries on the continent; the state relied heavily on foreign aid. Since the 1970s however, with the discovery of diamonds, Botswana began an aggressive and highly efficient government development policy of centrally managing the revenues from its mineral resources in a relatively benevolent and transparent manner. Botswana built strong institutions and policy frameworks that have turned what would have otherwise become a "resource curse" into a blessing.

To a larger extent, decision making is based on broader consultations, in-clusive participation and consensus rather than coercion, intimidation, bribery and decrees as seems to be the case in much of conflict prone Africa. Several scholars have therefore argued that it is the leadership's conscious effort to shape Botswana into what it is today – "a functioning democratic developmental state" that largely explains the success of this country. Much as Botswana still faces many challenges, it has manifested itself as a Democratic Developmental state as described in the literature and in the next paragraphs I try to briefly discuss what characterizes a developmental state.

According to Chalmers Johnson (1982) the four segments of a developmental state include; the presence of a small but professional and efficient state bureau-cracy; a political milieu where this bureaucracy has enough space to operate and take policy initiatives independent of intrusive interventions by vested interests; the crafting of methods of state intervention in the economy without sabotaging the market and a pilot organization such as Chalmers found in MITI[40]. Left-wich, on the other hand, arrived at some defining characteristics of a typical de-velopmental state, which according to him, comprise of six major components: relative autonomy; a powerful, competent and insulated bureaucracy; a weak and subordinated civil society; the effective management of non-state economic interests; and legitimacy and performance (Leftwich, 1995: 405).

Leftwich (1996) further describes how in a developmental state, the elite/leader's deliberately decide to work together to achieve a particular developmen-tal vision for their society. He argues that such elite often have the ability to manage conflicts, devise strategies for cooperation, negotiation and compromise which are designed to avoid alienation and disgruntlement among large sections of the society. This, as we shall see in the next sections, is what largely explains Botswana's success.

In a comprehensive review of South East Asian Developmental states, Ha-

---

40. The MITI is the Ministry of Trade and Industry in Japan that has directed that country's transformation since the 1940s.

Joon Chang argues that successful developmental states have pursued policies that co-ordinate investment plans and have a national development vision, implying that the state is an entrepreneurial agent. He argues that such states engage in institution building to promote growth and development and play a key role in the management and mediation of conflicts that arise out of reactions and counteractions to the development trajectory (Chang, 1999: 192-199). Chang stresses the building of strong state institutions as being a crucial factor of developmental states, similar to what have been characteristic of Botswana. The next section tries to give an explanation for Botswana's current success in light of the developmental state arguments discussed above.

b. The Ideology & Attitude of Botswana Elite from Independence

It has been argued that in Botswana, there has been a long and sustained commitment by the state to pursue national development. This goes back to the first presidency of Sir Seretse Khama (Parsons, Henderson and Tlou, 1995). A conscious and disciplined leadership in Botswana has seen, as one of its main duties, the need to develop professional institutions with competent bureaucrats. Indeed, the very process of post independence nation-building took on a nature that was inspired by the fundamental task of development at all levels of society and government. This developmental ethos was accepted and advanced by both the political and bureaucratic elites and by the institutions that they built up (Tsie, 1996). This echoes Ha-Joon Chang's argument that a developmental state should act as an entrepreneurial agent whilst engaging in institution and capacity building. In 1981 the then Minister of Finance and Development Planning, Peter Mmusi, spoke of the need for a 'purposeful government' which acquires the expertise to deal with companies on its own terms (Sebudubudu & Molutsi; 2011). Botswana has been able to put together strong negotiating teams, backed with well-worked-out negotiating mandates to oversee the implementation of major mining agreements with detailed care (Harvey and Lewis, 1990: 119). Attempting to account for how and why a disciplined and competent state apparatus emerged post-independence is what we shall turn to in the next paragraphs.

It is important to single out and understand the contribution and centrality of President Seretse Khama's leadership if any correct analysis of why Botswana has emerged a success story is to be made. As noted earlier, Sir Khama's strong leadership and early decisions had a great impact on the growth of the Botswana state and its institutions and proved to be decisive in establishing a strong state. As Acemoglu et al (2003) argues, President Khama's far sightedness in passing the 1967 mines and minerals act, vested the government with sub-soil mineral rights which has helped to avert any conflicts that would have otherwise emerged (notably if one tribe controlled the land in which the minerals were discovered).

Furthermore, Du Toit (1995) presents that there are several other factors that can be noted as key decisions by Khama which led to the strengthening of the Botswana state. These include among others, the separation of state personnel from politics, the transfer of land allocation from chiefs to the state, the incorporation of customary courts into the state legal system, the on-going presence of significant expatriate personnel in the public service and the co-option of potentially chiefly challenges (Du Toit 1995, cited in Philippe Martin 2008).

While it's true that Botswana's leaders and elite have played a critical role in the country's development, there are other contextual factors that have contributed to its success story that can offer lessons for Uganda. One rather interesting unique feature about Botswana is that at independence the British Protectorate had deliberately developed nascent political institutions that brought together modern and traditional elites as well as African and European leaders. This helped foster a common understanding among a cross-section of the elite that took over at independence. According to Sebudubu & Molutsi (2011), trust between key actors who inherited the post colonial state contributed to the forging of what were otherwise hostile relations between different elite in Botswana at the time. The major roles that key leaders like Sir Seretse Khama and Quett Masire played in elite coalition building is one factor that cannot be ignored in Botswana's development trajectory. Many analysts like Sebudubudu & Molutsi (2001) and Harvey (1992), acknowledge the exemplary leadership of the country's first president. Khama's exceptional integrity (pragmatism, inclusiveness, commitment to racial and social unity [married to a British white woman] and refusal of nepotism, established a precedent for high ethical standards and set the tune for his successors (Sebudubudu & Molutsi; 2001). It is under Khama's leadership that a strong and relatively independent and accountable merit-based civil service was created for which government pursued a developmental approach. A conscious decision was made to invest the proceeds from the diamond industry to the greater good of society and the nation at large. A political culture of self-restraint and accommodation was therefore institutionalized from independence.

Notably, Battistelli & Guichaoua (2011) present that Khama held a strategic position in Botswana's polity that conferred on him high esteem in the eyes of many because not only was he part of the emerging elite, but he was also a chief of the biggest tribe, the Bamangwato. This undoubtedly won him inherent legitimacy that other aspiring leaders could not measure up to. Indeed most of the leaders who took over at independence in Botswana were; as Battistelli & Guichaoua (2011) put it; *westernized traditional rulers* unlike the case in most other parts of Africa where it was mainly *westernized revengeful technocrats turned anti-colonial activists*, who took over state power.

The schoolmate friendship amongst the elite that took over leadership at independence also further explains much about the ease with which a "grand coalition" was successfully formed in Botswana to guarantee political, social and economic stability. Many of the leaders married their classmates or friends of their sisters, brothers and cousins, and invited each other to form political parties or community based organizations. They recruited one another into public service and even formed private businesses   collaboratively.

According to Sebudubu and Molutsi (2011), Botswana's elite is uniquely different in the sense that they are all predominantly from one tribe, the Tswana, and have a common cultural background, one language and a common cultural orientation. In addition, because many of the elite studied together in institutions outside the country or in a few elite schools within Botswana, they developed a common political and social value system. Arguably, education played a key factor and underpinned the coalitions that emerged later in Botswana's development Process. It was easy to build networking relationships. Botswana has had only one University since Independence until only very recently and this meant that generations of elite went through the same educational system, and institutional experience which helped pave the way for broad based inclusiveness with regards to leadership (Sebudubu & Molutsi, 2011). It is clear that Botswana utilized a system of inclusion and collaboration, while few could argue this has been the case throughout post-independent Africa.

Furthermore, most presidents at the time of Independence were teachers and with roots in socialism; Zambia (Kaunda), Tanzania (Nyerere) Ghana (Nkrumah), while the founders of Botswana's nationalist party like Seretse Khama and Kitumire Matsire were 'liberal minded' prominent cattle and land owners. Being historically associated with wealth accumulation and production, the Tswana elite have therefore had an interest in upholding a legal framework governing property rights and resolving commercial disputes (Sebudubu & Molutsi, 2011).

Botswana's success lies in its ability to devise a larger political strategy of balancing regional, ethnic and racial interests that have enabled Batswana elite to work together in harmony for a common developmental agenda that has transformed the country into a middle income economy. The next section looks at the process of building a grand coalition.

c. Building a Grand Coalition as a Strategy for Nation-Building in Botswana

Botswana is made up of a number of Tswana speaking groups which constitute the dominant social cultural and political pattern of that society. At independence, the new leaders of Botswana at the national level were mainly preoccupied with ensuring that leaders of the different ethnic and racial groups were included in the "grand coalition"[41]. The institutions of chieftaincy and of traditional assemblies and courts which were political, judicial and social pillars of pre-colonial and colonial Tswana society, still exist and are an integral part of modern day Botswana. It is therefore conceivable that here-in lies the model of ensuring stability and development for African countries. The fact that leaders in Botswana managed to successfully blend the county's traditional and modern institutions when in most other countries the traditional institutions have been abolished or transformed dramatically is something worth taking note of. A closer look into how to value and revive culture and its centrality in shaping thinking is one aspect that needs to be further explored and understood.

According to Sebudubudu & Molutsi (2011), the process of blending the two institutions in Botswana was neither easy nor smooth. It involved several struggles, resistance, resignations and carefully designed political strategy of coalition building and inclusiveness by the new elite. Nevertheless, the careful management of traditional chiefs and their institutions (i.e. giving them recognition and being accommodated and integrated into the new state) ensured continuity and arguably explains Botswana's unique post-colonial stability and enviable growth and development. Evidently, Botswana was able to devise the right strategies to defuse social tensions.

It is worth noting that of all the parties formed in Botswana at independence, it is the BDP (Botswana Democratic Party) that chose to use a more inclusive strategy of recruiting leaders from royal families who were either chiefs or had relations to chiefs. These were self-made emerging leaders like Masire who eventually came to be President and also leaders of the white settler communities among them. Sebudubudu and Molitsi (2011) argue that, unlike all the other parties in Botswana at Independence, BDP managed to form a grand coalition of strategically well placed leaders from the very beginning. The BDP was therefore able to make a successful appeal to the wider section of the population and won the first election in 1965 and has continued to win up to today. We now take a look at the institutions in Botswana.

d. Creating the Right Institutional Framework for Development in Botswana

Several scholars including Mehlum, More and Torvik (2006a) and Easterly and Levine (2003) argue that the quality of a country's institutions determines

---

41. The country's peculiar ethnic structure and the nature of its colonial history have impacted positively on Botswana's development trajectory. Unlike other African Countries, Botswana has no single dominant monarch or ethnic group. In effect, there was no dominant tribe big enough to influence any other ethnic tribes whether politically, socially or economically. The situation created an environment like the one in some parliaments elsewhere where there is a multiplicity of parties of different strengths in parliament and none enjoys an absolute majority. It was therefore inevitable for the first generations of the country's post-colonial leaders to generate an inclusive political strategy (Sebudubudu & Molutsi 2011).

its level of income per capita. They present that the main reason for diverging growth experiences of resource rich countries lies largely in the differences in the quality of institutions. Easterly and Levine (2003) utilize an institutions index that takes into account such factors like voice and accountability, political stability and absence of violence, government effectiveness, regulatory burden, rule of law and freedom from graft. Their main assertion is that institutional quality is a key determinant for a country's long term economic growth.

However, the most important flow of research focusing on the nexus between Institutions and natural resources has revolved around what could be termed *"the rentier state argument"*. Ross (1999) contends that when governments gain most of their revenues from external sources, such as resource rents or foreign assistance, they are freed from the need to levy domestic taxes and become less accountable to the societies they govern. Not only is accountability towards citizens discouraged, but embezzlement of collective wealth by the leaders is also very likely to happen. As earlier discussed, Botswana's institutions evolved from both traditional (pre-colonial) Tswana culture and British influence. The traditional system provided for consultations between the chief and his people; the chief in Tswana culture was responsible for looking after the provision of goods and services including law and order (Leith 2005). This meant that at Independence, when the traditional institutions were integrated into the political system, the political elite were well aware of the constraints on their rule and knew that they were and had to be accountable to their people. In addition, a tradition of public participation and consultation was embedded in the policy process as this had been the way of Tswana culture. According to Phillippe Martin (2008), these structural factors were to have a great influence in post-independence decision-making regarding the country's development and provided an environment conducive for growth-promoting policies. As one scholar has put it; relatively well institutionalized private property values, rule based governance and independent judiciary explain the good management that has taken place in Botswana and has contributed to encouraging the sustained inflow of foreign direct investment (Maipose, 2003). The next section looks at how Botswana has effectively managed its mineral wealth.

e. Botswana's Management of its Mineral Wealth

Botswana's management of its mineral resources is unique in the sense that at independence, the political leaders decided to pro-actively enter into a strategic alliance with international capital regarding ownership and management. Mineral rights were taken from the tribes and declared property of the state in 1967 (Battistelli & Guichaoua, 2011). The fact that the first minerals were discovered in Khama's territory, the Bangwato area, facilitated this process for the new leadership, yet it also lends much credit to his decision to renounce private gain by his tribe for the good of the nation.

The mining sector was developed through a smart partnership of strategic cooperation between the government and private sector where the government has shares in the company but leaves management to the private owners and is mainly interested in how dividends and taxes are used. The partnership that was created stands as one of the central pillars of Botswana's success and is reputed to be 'one of the best ever secured between a developing country government and a major multinational mining company (Jefferies 1998).

The state from the very beginning established both control of mineral rights but also entered into strategic partnerships with international companies. The leadership in Botswana was able to negotiate a uniquely strategic partnership with De Beers and form a distinctive successful coalition that has managed to mine, manage and share Botswana's mineral wealth in a manner that has benefitted the country as a whole. Botswana has realized its success from the joint effort and cooperation of external interests in the private sector. As Sebudubudu & Molutsi (2011) affirm, the private sector in the country's mineral sector carved out a unique strategic and enduring relationship with the Botswana State from early on; it is this unique, friendly relationship that has enabled the country to crucially avoid collusion, rent-seeking or predatory behavior. The Botswana Development Corporation (BDC) is the public company that represents and coordinates government activities in these ventures similar to what Chalmers (1982) described in his categorization of developmental states.

When diamonds were discovered in the mid-1960s, in the Bangwato territory of Orapa and Letlhakane, the Government entered into a uniquely successful partnership with De Beers which had for many years been mining diamonds in South Africa, Namibia and Angola. The international company accepted a uniquely generous agreement with the government of Botswana whereby the Government and de Beers agreed to a 50/50 split of the diamond mining revenues (Maipose, 2003). With the discovery of a new and even larger and more valuable Kimberlite deposit in Jwaneng in Bangwaketse territory in south Botswana in the late 1970s, the government negotiated for 60% revenue from the mine. Subsequently, a jointly owned company named De Beers – Botswana Mineral Company (or DEBSWANA) was formed with equal membership and a chairmanship of the board rotating between officials from the government of Botswana & De Beers (Sebudubudu & Molutsi 2011). Today it is DEBSWANA that runs the diamond mines in Botswana. Even more interestingly, the Government of Botswana in 2007 was able to compel the De Beers subsidiary company, Diamond Trading Company (DTC) originally based in London and dealing with the downstream elements of the industry including cutting and global marketing of diamonds, to relocate to Gaborone and open up its offices in

Botswana. This was an important accomplishment in negotiations for the government as it demonstrated their ability to successfully apply pressure to a large multi-national, and demonstrated how the government had an interest in ensuring De Beers integrated further into Botswana. By relocating offices, De Beers demonstrated they are committed to the success of the state (Sebudubudu & Molutsi 2011).

f. Economic Prudence of the Political Leadership & Centralized Planning

Scholars like Sebudube and Molutsi (2011) have argued that Botswana has been able to make advances in part, due to the economic prudence of the leaders at independence which prompted centralized planning. Apparently, one of the reasons Botswana developed a culture of economic prudence at independence was due to the tough economic situations it found itself in at the time.

Botswana gained independence in a period when the international community was not wholly averse to state involvement in the economy and as such, the leadership chose to pursue a strong central role in formulating, managing and administering aspects of economic change from Gaborone. The fact that the country was very poor, and relied heavily on South African aid, made it very vulnerable. The BDP pushed for adoption of careful and pragmatic policies with a preference to centralized planning (Battistelli & Guichaoua; 2011). This approach has never been relinquished by the state. The Ministry of Finance & Development Planning in Botswana today exerts extensive powers with respect to budgetary and financial affairs of the government and oversees all major decisions on economic growth and national plans. The rationale adopted by government at independence was such that since the nation lacked natural resources (diamonds were not yet a part of the economy), careful centralized planning would be required to put the limited resources to good use (Siphambe, 2007).

Botswana has used what has become known as rolling plans with regards to economic management, which are reviewed and updated on a six year basis. The level of public/private consultation as well as the level of strict rules governing the process has resulted in a very stable and effective system of national fiscal management. (Beaulier and Subrick, 2007)

g. Land Governance in Botswana & Issues of Public Private Sector Coalition

Another key action taken by the leaders of Botswana at independence that did go a long way in shaping the development of the country was the transfer of land ownership and control from the tribes to the central government. Legislation to that effect was introduced and the new Botswana state was very clear that land was a key resource which needed to be released from the control of chiefs and private individuals in order to be given to citizens to develop and improve the country's agriculture and, where necessary, for use of the government for development projects.

On the issue of public and private sector coalition, the political elite in Botswana has also showed its commitment to attracting Foreign Direct Investment (FDI) not only by entering into partnerships with companies like De Beers but also ensuring that they create a conducive environment for the private sector. The government has gone out of its way to ensure a stable macro-economic and political environment that has attracted many FDIs. Company tax in Botswana for instance is one of lowest in the world at 12% (Sebudubudu & Molutsi 2011). The government also introduced a number of business initiatives in the 1980s and 1990s to finance both local and foreign business development in the country. The Financial Assistance Policy (FAP) and its successor, the Citizen Entrepreneur Development Agency – (CEDA), have been established to boost private investment in different sectors of the economy which has seen the economy flourish.

Another factor worth noting is that of consultative decision making. The Government has institutionalized policy ownership between the state and private sector by ensuring that there is constant consultation between government and the private sector. The government discusses development issues, policies and strategies with the private sector.

h. The Role of Foreign Technical Assistance and Civil Society

One other aspect of Botswana's success worth noting is the fact that localization was not rushed and introduced at the expense of merit. While other African countries, including Uganda, rushed to Africanize their civil service at independence, Botswana decided to retain most of its expatriates who played a key role in the economy by providing much needed technical expertise (Parsons, Henderson & Tlou 1995). Senior Positions in the public service and parastatal organizations were occupied by persons of different ethnic and racial groupings over the years and qualified individuals from other countries were also recruited until only recently. To some degree, this gradual replacement process ensured the public service remained sufficiently competent until the technical skills of the locals were built up (Battistelli & Guichaoua; 2011).

On the role of civil society, analysts like Molutsi and Holm, (1990) have argued that because the state has curtailed activities of civil society in the past, civil society in Botswana is seemingly weak casting doubts on Botswana's democracy. However there are also observers arguing that in fact civil society in Botswana acts differently and prefers to engage in a non-confrontational approach with government through negotiation. Political and Social discontent tends to be mitigated by finding compromises which satisfy different constitutions without conceding the space that could challenge the ultimate authority of the state. The absence of experiences with popular struggles has lent society a less volatile and exploitative experience as seen elsewhere. This is a unique approach that Uganda could perhaps learn from.

Similarly, the same political culture of non-violent negotiation is detectable on the government's side. Very few cases of state sponsored violence or abuse have taken place. The system in Botswana has arguably been responsive and has tried to accommodate the concerns of those affected by measures seen necessary to maintain centralized control. This offers lessons for many countries in Africa.

In conclusion to this section therefore, Botswana's success can largely be traced to the evolution of a genuinely conscious elite leadership which has worked though consultation, consensus building and inclusive strategies to drive successful development. It has been the leader's conscious effort to create a particular type of politics and state that has made Botswana into what it is today.

The judicious balance between the various ethnic groups has also been a key factor and has ensured that no single group has dominated the political space. Both at independence up to today, all tribes participate in national affairs.

The mineral coalition that was crafted and the relative good use of the mineral proceeds for the benefit of all, has enabled the country to remain peaceful and ensure the continued steady economic growth. While Botswana is unique, there are some lessons that can be learned and applied to the case of Uganda and this is what we turn to, in the next section.

In general, the Botswana model suggests that a form of *'state developmental management'* of a natural resource is indeed feasible, but this has to be of the right kind. Leftwitch's (2000) argument is of particular relevance here; i.e. that it is not the *'amount of state involvement'* in the market that matters, but rather the 'kind' of involvement. But what has the situation in Uganda been? What is the nature & character of Uganda's leadership and how has the government managed to balance ethnic, racial and regional interests in Uganda in light of the emerging oil industry? This section tries to discuss major lessons that Uganda can draw from the Botswana experience and the policy implications following from the empirical insights discussed throughout the paper.

One major lesson to learn from the Botswana example is the need to avoid personalisation of power (see Box 1) and build strong state institutions to govern society. In Uganda, more still needs to be done to avert this. State power appears to be increasingly becoming personalized and because of endemic corruption, most of the institutions previously created to ensure effective management of the economy are arguably not functioning as would be expected. The government would do well to look to Botswana's example of benevolent leadership and crea-

---

Box 1

President Museveni appointed his wife, Mrs Janet Museveni, as Cabinet Minister for Karamoja; his brother, Gen. Salim Saleh, formerly a minister of state for micro finance, as Senior Presidential Advisor on defence, a job at the same rank as a cabinet minister; his brother-in-law, Sam Kutesa, minister of foreign affairs; his son, Muhozi Keinerugaba, commander of the Special Forces, his daughter Natasha Karugire, Private Secretary to the president in charge of Household. President Museveni also appointed his nephew, Joseph Ekwau (son of his younger sister Violet Kajubiri), Private Secretary to the President in charge of Medical Services (HIV//AIDS); his sister Miriam Karugaba is Administrator at State House (she is semi-literate) and her husband (therefore Museveni's brother-in-law), Jimmy Karugaba, is Officer in Charge (OC) of the Accounts Department at State House. Museveni has also appointed his sister-in-law, Jolly Sabune, Executive Director of Cotton Development Authority, his niece-in-law, Hope Nyakairu, Undersecretary for Administration and Finance at State House, his cousin Bright Rwamirama, State Minister for Animal Husbandry and the list goes on and on. Many observers say that increasing family influence in government has gone hand in hand with the informalisation of power. (Source: The Independent Magazine, Family Rule in Uganda, Wednesday, 11 March 2009)

---

tion of strong institutions and invest more efforts in promoting clean leadership and strengthening the governance institutions. If Uganda can put more efforts in combating corruption and strengthening its public institutions, it would be able to control the economic influences of oil. The leaders especially need to change the rules of the game and begin to act differently to avoid the grabbing of oil revenues via corrupt practices. Countries that have successfully transferred from corrupt to less corrupt systems seem to share the characteristic that actors at the very top of the system i.e the public officials at the high level, have served as role models[42]. The modus operandi for leaders in Uganda needs to change fundamentally.

Another lesson learnt from Botswana's example is the need to build social cohesion and create a common vision to transform society, a strategy that President Khama used and largely explains Botswana's success. One could argue that President Museveni is perhaps recruiting mainly people from Western Uganda in the top positions of government, with examples in the Text Box 1 to try and achieve this. Perhaps his idea has all along been to create a caliber of leaders that can easily trust and speak to each other and develop a common political & social value system for the country. Many observers however note that this is nothing other than systemic nepotism. In a highly ethnically divided country like Uganda, with no historical connections of co-existence, such a project would not be feasible especially when seen in light of the key principles of a Democratic Developmental State. Throughout Uganda's history, the nation has experienced civil strife, internal conflict and numerous coup attempts. Grievances across the country remain high especially between the government and ethnic groups who feel they have been largely neglected. Oil has increased national expectations largely due to the rhetoric used by President Museveni. However, as national expectations rise, group grievances still exist. Due to its highly fractionalized state, Uganda's leaders from the time of independence, and up to today, would have done and could do well to scale up a model similar to what was developed by Botswana's post-colonial leaders.

In line with Francis Fukuyama's arguments about trust[43], it is no surprise that Uganda which is a "low trust" country, has somewhat failed to form a

---

42. In Hong Kong and Singapore for example, corruption was successfully fought from "above" implying that the members of the ruling elite themselves set an example by changing their behavior beyond rhetorical level (Root, 1996)

43. Francis Fukuyama (1995) in his book; *'Trust; The Social Virtues and The Creation of Prosperity"* examines a wide range of national cultures in order to describe the underlying principles that foster social and economic prosperity. Insisting that we cannot divorce economic life from cultural life, he contends that in an era when social capital may be as important as physical capital, only those societies with a high degree of social trust are able to create the flexible, large-scale business organizations that are needed to compete in today's global political economy.

broad coalition like the one in Botswana. Social cohesion around the right policy framework to transform the economy and most especially on the emerging oil
industry[44] has failed to form in Uganda. Although the NRM did for sometime
after the bush war attempt to build social cohesion in much of the country, this
seems to have arguably dissipated with the revival of multi-party democracy in
1996 and the increasing personalization of power by President Museveni. One
reason Museveni has arguably ended up with so many relatives in key positions
is because he has limited the independent growth of state institutions in Uganda
(See Text Box 2). It is very difficult to govern without organized institutions unless one has a force to rely on to counter challenges to ones authority. Perhaps
that is why the security forces have became the bedrock of President Museveni's
power[45]. What we see playing out on Uganda's political scene today is a myriad
assortment of loose factions with unclear common objectives and no underpin-
ning social values as seen in Botswana. Most coalitions work as patronage groups
to the regime in power. Such loose factional leadership networks like the *"Young
Parliamentarians"* and *"NRM caucus"* cannot form any *"Grand Coalition"* as was
the case in Botswana because of the ethnic conflicts, intrigue and infightings
which continuously destabilize government and hinder development. Uganda
is in a unique position to learn from the mistakes of its neighbors and righting
all the wrongs, more so, building a national coalition to transform the country.

Another lesson from the Botswana case is that there is need to devise the
right strategies to diffuse social tensions. Botswana adopted an all- inclusive
strategy to accommodate traditional institutions in the political setup after independence and this enabled them build a grand-national consensus. The mate-
rial circumstances of the country and the implications of this for its success as
an independent state were crucial in promoting momentum for the formation
of a grand coalition at political, public and economic levels. Uganda needs to
devise strategies to accommodate the differing factions in order to build social
cohesion. Although Ugandans might have a shared desire to see the country
grow and prosper, the reality of a scattered and rural population means that
many may feel remote from the process of decision making. President Musev-

---

44. In October 2011, allegations of rampant corruption in Uganda's oil sector, in a moment of
rare bipartisanship, ended in the suspension of new deals between Uganda and foreign oil
companies and led to the resignation of some ministers signaling an increasingly assertive
parliament ready to challenge the regime against corruption. Of recent however, in January
2013, President Museveni, Defense Minister Crispus Kiyonga and Chief of Defense Forces
Aronda Nyakairima all warned that the military could intervene to "refocus the country's
future" if the current "bad politics" in parliament continue. This arguably signals that he is
beginning to feel the heat and possibly threat to his power. (For more see Article by Ruth
Peluse in *Think Africa Press*, 2013)
45. See: Think Africa Press (2012): 'How was Museveni stayed in power? http://thinkafrica-
press.com/uganda/editor-q-a-how-has-museveni-stayed-power

eni, as well as Uganda's other civil servants have yet to demonstrate the level of prudence employed by Botswana's leadership. While many Ugandans have access to very limited social services, the government routinely mismanages public finances for personal profit.

Furthermore, although Uganda has improved its processes of formulating plans and has tried to implement a decentralization strategy, the country is still fraught with a lot of challenges. Different from Botswana, the lack of capacity in local governments has resulted in poor service delivery and wide-spread allegations of corruption. Progress in dealing with corruption is urgently needed. In addition, the executive seems to have an upper hand in making decisions at times even by-passing parliament. No wonder the management of the emerging oil industry has raised a lot of debate especially with the recently passed petroleum bill in parliament which gave a lot of powers to the Minister of Energy. This was arguably contrary to the perceptions of many Ugandans. While the government has a number of robust formal structures in many aspects, decision making often bypasses official channels. The purchase of fighter jets in 2011 – costing some USD $ 740 million withdrawn from the Central Bank without prior approval is one such case[46]. The risk is that resource falls from the oil sector could also be used in this pattern with little public consultation and this may precipitate a resource curse. The executive has on many occasions exercised strong influence over key policy areas relating to oil. This approach will need to be reversed if Uganda wants to avoid the resource curse.

The Botswana case also shows how the government was able to manage land governance challenges right from independence. In Uganda however, land issues are still a challenge. Land ownership for instance, especially in the oil rich Albertine region has been fraught with a lot of controversy due to the varying land tenure systems and failure to manage conflicts effectively due to weak institutions. Lately, all kinds of actors are taking advantage of the corrupt institutional framework in the country to displace innocent civilians from their land without adequate compensation.

As discussed in the paper, one of the biggest threats to Uganda's emerging oil industry that the government needs to address is the lack of transparency. Allowing rumours and speculation to dominate the news and media can easily create huge disharmony amongst the population especially given the latent divisions along ethnic, religious and regional lines. As noted in the first section of the paper, transparency is vital. With accurate and reliable information, the government can turn the negative perceptions into optimism. Uganda can learn from countries like Ghana which has put in place a legally constituted Public Interest Accountability Committee which brings together representatives of aca-

---

46. How Museveni convinced MPs on Fighter Jets: *The Observer*. 7th April 2011

demia, NGOs, churches and traditional authorities to monitor and report on the oil sector.

In addition, the Ugandan state has had numerous challenges in dealing with divergent views to government. Clashes between civil society and government are very common and on the increase and these curtail harmony and social cohesion. The government has increasingly put civil society under pressure, particularly those organizations that appear to be infringing upon the officials' political and financial interests. Research and advocacy organizations in Uganda that deal with controversial topics are facing increasing harassment by the government (Human Rights Watch, 2011) Groups have recently experienced forced closure of meetings, threats, harassment, arrest, and punitive bureaucratic interference. The government needs to change its approach and improve the operating space of civil society in order to build social cohesion. There is need to ensure that strong social actors emerge instead of considering any divergent view as opposition to government. There is need to provide space for the population to articulate alternative views and perspectives on the overall direction of the country, in many ways similar to what Botswana has been able to put in place. The government needs to listen to local voices especially with regard to spending priorities of the oil rents.

Furthermore, Botswana was able to integrate its traditional rulers in the main stream decision making process yet Uganda has failed to do this. Traditional rulers in Uganda are undoubtedly extremely important and command both loyalty and respect. Unfortunately, their role in politics is rather limited and indirect, both by law and custom. Churches and religious leaders are also important but divided along regional and political lines. For Uganda to develop strong social voices that will enable the creation of a social cohesion which is necessary to manage natural resource effectively, civil society and especially the role of traditional leaders *(like the King of Bunyoro who presides over the oil region)*, will have to be strengthened to play a vital role in the management of Uganda's oil.

Finally, although Uganda has taken steps to create competent capacity in the field with a number of qualified specialists in the oil industry, a lot more needs to be done. A competent bureaucracy is one of the factors that explains Botswana's success and should be adhered to by Uganda. Due to the vibrant debate in the media, many Ugandans have been able to learn about oil and the NGO sector is well informed. The establishment of oil training facilities most importantly Uganda Petroleum Institute is worthy of note and a positive step in building technical skills and ensuring that oil related jobs are taken by Ugandans. The focus however has been on the lower cadre technicians and training of more technical specialists at the higher levels is most urgently required.

All through the paper, I have argued that although the literature, general perceptions of Ugandans and the news in the media suggest that oil will lead Uganda down a slippery slope, the state actually has the opportunity to learn from a country like Botswana and avoid the curse. Uganda has relative strengths that must be recognised which can offer a sound foundation for meeting the challenges of oil although there are many issues of concern too. I argue this based on Uganda's remarkable turnaround, from a post-independence catastrophe, to one of Africa's relatively success stories, despite the many challenges still.

The paper explored the ongoing debates in development literature and within the media, analyzing all relevant oil related articles in the Daily monitor and New Vision from October 2010 to much of the present and tried to assess people's perceptions about the oil industry in Uganda. Based on this analysis, the paper concludes that scepticism still exists and will most likely continue unless the government puts in place mechanisms to manage the oil industry in a transparent and more efficient manner with the right institutional frameworks. Based on the combination of the failures in Africa, increasing corruption in the county, the existence of weak institutions and lack of transparency on how the oil sector operates, I argued that scepticism in Uganda exists and is reasonable. I argue that Ugandans as well as institutions such as the World Bank and International Alert are sceptical of oil based on its impact on economics, institutions and on civil conflict.

Yet although every African oil state to-date has been unable to prevent the onset of a resource curse, this does not mean Uganda is destined for a similar failure. I argued that due to a number of factors like Uganda's relatively improving democracy, its ability to fight HIV/AIDS, it ability to turn the post-Amin and post-Obote catastrophes into success and the fact that it can learn from success countries like Botswana, Uganda may not be destined to follow the path set by other African oil states. I try to provide a clear explanation of what explains Botswana's success and what lessons Uganda can learn from this. I argue that if President Museveni and the NRM wish to develop Uganda, it is prudent of them to do all they can and borrow lessons from a success story like Botswana to prevent the resource curse.

However, I also noted that the recent corruption trends do provide cause for concern. For Uganda to transform itself into an inclusive success especially with its burgeoning oil industry, I argue that there is an urgent need to avoid systemic corruption and nepotism. Uganda should look to strengthen state institutions while developing a system of cross-party collaboration and build social cohesion towards a common vision for prosperity. A firm interest in the success of public private partnerships as well as a well-crafted approach to managing traditional and modern institutions would substantially aid Uganda in its development agenda.

Uganda is in a unique and fortunate position. It has the opportunity to learn from the mistakes made by other African oil producing countries and has the knowledge available to it on how to do so. Uganda is undoubtedly at a crossroads, what it does with its oil reserves will have a profound impact on its future. As argued throughout the paper, for Uganda to take full advantage of its oil reserves and become Africa's first oil success story, it must combat corruption, become more transparent and accountable, strengthen public institutions, make decisions with public interest in mind and build social cohesion. Although all of Africa's oil producing nations have largely been unable to avoid the curse, this paper suggests that we cannot immediately or easily write Uganda off as a lost cause. By highlighting current issue areas and objectively assessing Uganda's emerging oil industry to date, this paper suggests that we should not write off Uganda immediately but rather adopt the strategy of cautious optimism.

References

Acemoglu, D., S. Johnson, and J. A. Robinson (2003). "An African Success: Botswana." In *Search of Prosperity: Analytical Narratives on Economic Growth*, ed. D. Rodrik. Princeton, NJ: Princeton University Press.

—, and J. A. Robinson (1999). "On the Political Economy of Institutions and Development." *American Economic Review* 91 (4): 938–63.

Acquah, B. K (2005): "National Development Plans in Botswana" in Siphambe, H.K, Narayana, N. Akinkugbe, o. and Sentsho, J. (eds) *Economic Development of Botswana: Facet, Policies, Problems and Prospects*. Gaborone: Bay Publishing.

All Africa Press: 19th Feb 2013; 'Uganda: Museveni's Muzzled MPs - Uganda's Missed Opportunity in the Oil Debates' Retrieved from: http://allafrica.com/stories/201302200617.html?page=3

All Africa Press: 13th Nov 2012; 'Uganda Legislators Clash Over Minister's Oil Powers' Retrieved from: http://allafrica.com/view/group/main/main/id/00020853.html

Bates, R.H (2008). *When Things Fell Apart: State failure in Late – Century Africa*. Cambridge: Cambridge University Press.

Battistelli S & Yvan Guichaoua, (2012) "Diamonds for Development? Querying Botswana's Success Story," in Thorp R, Stefania Battistelli, Yvan Guicaoua, Jose Carlos Orihuela & Maritza Paredes (2012) *The Development Challenges of Mining and Oil; Lessons from Africa and Latin America*, Palgrave Macmillan, NewYork.

Beaulier, S.A and Subrick, J. R. (2007): *Mining Institutional Quality: How Botswana Escaped the Natural Resource Curse'*, http:/www.scottbeaulier.com/REVISEDCURSE2__2_.doc. (Accessed Feb 2013)

Brunnschweiler, Christa N. and Erwin H. Bulte (2008). Natural Resources and Violent Conflict: Resource Abundance, Dependence and the Onset of Civil Wars. *Economics Working Paper Series* 08/78. Zurich: ETH.

Chang H-J (1999) 'The Economic theory of the Developmental State' in Woo-Cuming, Meredith (ed.) (1999) *The Developmental State*, New York: Cornell University Press, 182–199.

Collier Paul (2007): *The Bottom Billion*, New York: Oxford University Press.

— and Hoeffler, A. (2004): Greed and Grievance in Civil War; *Oxford Economic Papers* 56(4): 563-95

Du Toit, Pierre (1995): *State Building and Democracy in Southern Africa: Botswana, Zimbabwe and South Africa*. Washington DC: United States Institute for Peace.

Easterly William & Ross Levine (2003) "Tropics, Germs and Crops: How Endowments Influence Economic Development; *Journal of Monetary Economics* 50:3–39

Evans P. (1995), *Embedded Autonomy: States and Industrial Transformation* Princeton: Princeton University Press.

Fearon James (2005) 'Primary Commodities Exports and Civil War', *Journal of Conflict Resolution* 49(4)

Fukuyama, Francis (1995) *Trust; The Social Virtues and The Creation of Prosperity*. New York: The Free Press.

Ghazvinian, John (2007): *UNTAPPED: The Scramble for Africa's Oil*. Harcourt, Inc: Orlando.

Harvey C (1992) 'Botswana: Is the Economic Miracle Over?,' *Journal of African Economies* 1 (3).

Hillbom, Ellen. 2008. "Diamonds or development? A Structural assessment of Botswana's forty years of success.", *Journal of Modern African Studies* 46 (2):191–214.

Holm, John D. 2007. "Diamonds and Distorted Development in Botswana." Center for Strategic and International Studies, *Africa Policy Forum*. http://forums.csis.org/africa/?p=20

Humphreys, Macartan. 2005. "Natural Resources, Conflict, and Conflict Resolution: Uncovering the Mechanisms." *Journal of Conflict Resolution* 49(4):508-537.

—, Jeffrey D. Sachs and Joseph E. Stiglitz (eds). 2007. *Escaping the Resource Curse*. New York: Columbia University Press.

Jefferis, K. (1998); 'Botswana and Diamond-Dependent Development' In Edge, W. A. and Lekorwe, M.H (eds)

Botswana: Politics and Society. Pretoria: J. L. Van Schaik Publishers

Johnson, C. (1982) *MITI and the Japanese Miracle* Stanford: Stanford University Press.

—. (1993) 'Comparative Capitalism: The Japanese Difference', *California Management Review*, Summer.

—. (1999) 'The Developmental State: Odyssey of a Concept' in Woo-Cuming, Meredith (ed.) (1999) *The Developmental State*, New York: Cornell University Press, 32–60.

Juuko, Sylvia and James Odomel, 2008: Uganda: Country Will Not Export Crude Oil, *New Vision*, 16 June.

Kangave, J. (2006). Improving Tax Administration: A Case Study of the Uganda Revenue Authority. *Journal of African Law*, 49(2)

Karl, Terry Lynn (1977). *The Paradox of Plenty: Oil Booms and Petro States*. Berkeley: University of California Press.

Kizza Julius, Lawrence Bategeka, & Sarah Ssewanyana (2011); Righting Resource-Curse Wrongs in Uganda: The Case of Oil Discovery and the Management of Popular Expectations, *Research Series 78*. Economic Policy Research Centre, Kampala.

Le Billon, Philippe. (2001). "The political ecology of war: Natural resources and armed conflicts." *Political Geography* 20(5):561–584.

—. (2005). "The resource curse." *Adelphi Papers* 45(373):11–27.

—. (2008). "Diamond Wars? Conflict Diamonds and Geographies of Resource Wars.", *Annals of the Association of American Geographers* 98(2):345-372.

Leftwitch A. (1996) *Democracy & Development,* Cambridge: Polity Press

—. (1995) Bringing Politics Back In: Towards a Model of the Developmental State, *Journal of Development Studies* 31(3): 400–27.

—. (2000) *States of Development: On the Primacy of Politics in Development;* Cambridge: Polity Press.

Leith, J. Clark. 2005. *Why Botswana Prospered.* Montreal: McGill-Queen's University Press.

Maipose Gervase (2003): Economic Development and the Role of the State in Botswana' *DPMN Bulletin* X(2).

—, (2009) Botswana: The African Success Story, in *The Politics of Aid: African Strategies for Dealing with Donors.* Lindsay Whitfield, ed., pp. 108–130. Oxford, UK: Oxford University Press.

Martin Philippe (2008) A Closer Look at Botswana's Development; The Role of Institutions, *Paterson Review* – Vol 9. Ottawa

—. (2008): A Closer Look at Botswana's Development: The Role of Institutions in *Paterson Review* Volume 9.

Masire K (2006) *Very Brave or Very Foolish? Memoirs of an African Democrat* (edited by Stephen R Lewis) Basingstroke: Palgrave Macmillan.

Mbabazi P and I. Taylor (eds) 'The Potentiality of Developmental States in Africa: Botswana & Uganda Compared, Dakar, CODESRIA.

Mehlum, Halvor, Karl Moene and Ragner Torvik (2006) "Institutions and the Resource Curse" The Economic Journal 116 (508): 1–20

Molutsi PP (1989) *The Ruling Class and Democracy in Botswana* in J. Holm J & Molutsi P (eds) *Democracy in Botswana.* Gaborone: Macmillan Botswana Publishing Company (Pty) Ltd.

—. and J. Holm (1990) 'Developing Democracy When Civil Society is Weak: The Case of Botswana', *African Affairs,* 89 356.

Mwenda, A. (2007). Personalizing Power in Uganda. Journal of Democracy, 18(3), 23–37. DOI: 10.1353/jod.2007.0048

Önis, Z. (1991) 'The Logic of the Developmental State', *Comparative Politics,* 24 1, October.

Oredein Obafemi (2012),"Nigeria's Petroleum Industry Bill Could Become Law By Mid-2013 "http://www.epmag.com/Technology-Regulations/Nigerias-Petroleum-Industry-Bill-Become-Law-Mid-2013_109986 (Accessed March 2013.)

Parsons N, W Henderson and T. Tlou (1995) *Seretse Khama 1921 – 1980,* Gaborone, Botswana Society.

Robinson, John Alan, and Q. Neil Parsons. 2006. "State Formation and Governance in Botswana." *Journal of African Economies* 15 (AERC Supplement 1): 100–140.

—, Ragnar Torvik, and Thierry A. Verdier. 2006. "Political foundations of the resource curse." *Journal of Development Economics* 79 (2):447–468.

Ron J (2005) "Paradigm in Distress?" Primary Commodities and Civil War, *Journal of Conflict resolution* 49(4):443–50

Root, Hilton L. (1996) *Small Countries, Big Lessons: Governance and the Rise of East Asia*. Hong Kong: Oxford University Press.

Ross, Micheal L. (2012) *The Oil Curse: How Petroleum Wealth Shapes the Development of Nations*. Princeton: Princeton University Press.

—. (1999) 'The Political Economy of the Resource Curse'. World Politics 51 (2).

Rubongoya, B. Joshua (2007) *Regime Hegemony in Museveni's Uganda: Pax Musevenica*. Pelgrave Macmillan: New York.

Reuters News Agency, (2010). "Uganda to start building oil refinery in 2012". Retrieved from http://www.reuters.com/article/2010/11/30/uganda-refinery-idUSLDE6AT0O620101130

Sebudubudu D (2005) "The institutional Framework of the Developmental State in Botswana" in Mbabazi P and I. Taylor (eds) 'The Potentiality of Developmental States in Africa: Botswana & Uganda *Compared*, Dakar, CODESRIA.

Sebudubudu D. & Patrick Molutsi (2011): *The Elite as a Critical Factor in National Development; The Case of Botswana*, Nordiska Afrikainstitutet, Uppsala.

Shaxon Nicholas (2008): *Poisoned Wells: The Dirty Politics of African Oil*. Palgrave: New York.

Siphambe, H. K. (2007) 'Growth and Employment Dynamics in Botswana: A Case Study of Policy Coherence' *Working Paper No. 82*, Geneva: ILO.

Stevens P (2013) Resource impact – A curse or a blessing?: A literature survey. CEPMLP Electronic Journal – www.cepmlp.org (Accessed Feb, 2013)

The Daily Monitor, Feb 1st 2011, 'Tullow Oil exposes fallacy of investors'. Retrieved from http://www.monitor.co.ug/OpEd/Editorial/-/689360/1099094/-/91po7o/-/index.html (Accessed Dec, 2012)

The Daily Monitor, Dec 29th 2010; 'Lessons from Ghana's Oil journey'. Retrieved from http://www.monitor.co.ug/News/National/-/688334/1080222/-/cjfkatz/-/index.html (Accessed Dec, 2012)

The Daily Monitor, March 15th 2011, ' Rising Inflation hurting Ugandans standards of living'. Retrieved from http://www.monitor.co.ug/Business/Business+Power/-/688616/1124682/-/o55tqlz/-/index.html

The Daily Monitor, March 4, 2011; Oil and Gas can be a blessing for this country. Retrieved from http://www.monitor.co.ug/OpEd/Commentary/-/689364/1118908/-/13coff0z/-/index.html

The Daily Monitor, Nov 22nd 2010, 'PPP to shield Uganda from looming oil chaos'. Retrieved from http://www.monitor.co.ug/SpecialReports/Elections/-/859108/1057960/-/jli7a7/-/index.html

The Daily Monitor, Feb. 14th 2011. 'RDCs on the spotlight over oil'. Retrieved from http://www.monitor.co.ug/News/National/-/688334/1106768/-/c55urkz/-/index.html (Accessed Dec, 2012)

The Daily Monitor, Feb 4th 2011; 'Suspending award of oil licenses good move'. Retrieved from http://www.monitor.co.ug/OpEd/Editorial/-/689360/1101128/-/9f40xl/-/index.html (Accessed Dec, 2012)

The Daily Monitor, Feb 13th 2011; "Shall We Survive the Oil Curse?" Retrieved from http://www.monitor.co.ug/News/National/-/688334/1106380/-/c55xuuz/-/index.html (Accessed Dec, 2012)

The Daily Monitor, Feb 28th 2011; "Inflation projected at 7 percent by June as oil prices continue to surge". Retrieved from www.monitor.co.ug/business/commodities (Accessed Dec, 2012)

The Daily Monitor. Feb 6th 2011; 'When are the seeds of a curse sown?' Retrieved from http://www.monitor.co.ug/News/National/-/688334/1102060/-/c58sjoz/-/index.html (Accessed Dec, 2012)

The Daily Monitor, Dec 18th 2010; "Where are Ugandans in the lucrative oil deals? Retrieved from http://www.monitor.co.ug/News/National/-/688334/1074308/-/cjyb6lz/-/index.html (Accessed Dec, 2012)

The Daily Monitor, Jan 1st 2011; 'Hoima worries over oil land and pay..' Retrieved from http://www.monitor.co.ug/SpecialReports/-/688342/1081414/-/ubyl3t/-/index.html (Accessed Dec, 2012)

The Daily Monitor, Jan 17th 2011; 'Museveni: NRM has better plans to manage oil'. Retrieved from http://www.monitor.co.ug/News/National/-/688334/1090946/-/cisuufz/-/index.html (Accessed Dec, 2012)

The Daily Monitor. Nov 11th 2010; 'First family "too close" to oil sector. Retrieved from http://www.monitor.co.ug/News/National/-/688334/1051166/-/cl9icez/-/index.html (Accessed Dec, 2012)

The Daily Monitor. Nov 12th 2010; 'State House: Enemies fighting President on oil'. Retrieved from http://www.monitor.co.ug/News/National/-/688334/1051966/-/cl9c2qz/-/index.html (Accessed Dec, 2012)

The Daily Monitor. Dec 1st 2010; 'Mafias grabbing oil-rich land in Buliisa', Says MP. Retrieved from http://www.monitor.co.ug/News/National/-/688334/1063374/-/cklj80z/-/index.html (Accessed Dec, 2012)

The Daily Monitor. Dec 10th 2010; 'Uganda: Dutch government cuts Sh 10 billion aid over CHOGM scandal'. All Africa News Agency. Retrieved from http://allafrica.com/stories/201012100101.html

The Daily Monitor. Jan 14th 2011; 'Oil dilemma awaits as Museveni visits Bunyoro'. Retrieved from http://www.monitor.co.ug/News/National/-/688334/1089500/-/cj994qz/-/index.html (Accessed Dec, 2012)

The Daily Monitor, Dec 11th 2010; 'Mbabazi and Onek benefited from sale of oil production rights' Retrieved from http://www.monitor.co.ug/News/National/-/688334/1070248/-/ck2457z/-/index.html

The Daily Monitor March 16th 2011; 'I am not involved in oil deals, says Kutesa'. Retrieved from http://www.monitor.co.ug/News/National/-/688334/1125768/-/c3xhb4z/-/index.html (Accessed Dec, 2012)

The Daily Monitor, October 25th 2010; 'Uganda's next test is oil governance' Retrieved from http://www.monitor.co.ug/OpEd/Commentary/-/689364/1039312/-/13uiipyz/-/index.html

The Daily Monitor, Nov 22nd 2010; "Slippery politics of Uganda's oil". Retrieved from http://www.monitor.co.ug/SpecialReports/Elections/-/859108/1057918/-/jli760/-/index.html

The Daily Monitor, Feb 15th 2011; 'Uganda's oil age and the foreign exchange market'. Retrieved from http://www.monitor.co.ug/Business/Business+Power/-/688616/1107428/-/o6bvarz/-/index.html

The Daily Monitor, Feb 23rd 2013: 'Oil Bill Paves way for refinery'. Retrieved from: http://www.monitor.co.ug/News/National/oil-Bill-paves-way-for-refinery-bidding--says-minister/-/688334/1700988/-/fbybcyz/-/index.html

The Daily Monitor, 25th Nov 2012; 'What lies beneath the corruption in the Office of the Prime Minister?'Retrieved from; http://www.monitor.co.ug/OpEd/Commentary/What-lies-beneath-the-corruption-in-Office-of-the-Prime-Minister/-/689364/1628114/-/tny7dxz/-/index.html

The Daily Monitor, Dec 16th 2012: 'OPM theft outrage and debate on oil is a vote of no confidence in government' Retrievedfrom: http://www.monitor.co.ug/Magazines/ThoughtIdeas/-/689844/1644398/-/dt39lg/-/index.html

The Daily Monitor, Jan 3rd 2013: '2013: What will become of Uganda's oil fortune?' Retrieved from; http://www.monitor.co.ug/OpEd/OpEdColumnists/KaroliSsemogerere/2013--What-will-become-of-Uganda-s-oil-fortune-/-/878682/1656158/-/s0y1v4/-/index.html

The Independent Magazine; March 11th 2009; 'Family Rule in Uganda' Retrieved from http://www.independent.co.ug/index.php/cover-story/cover-story/690-family-rule-in-uganda

The Independent Newspaper 24th Sept. 2012; 'Museveni, Investors fight over refinery'. Retrieved from http://www.independent.co.ug/cover-story/6494-museveni-investors-fight-over-refinery-

The New Vision, February 17th 2011 - "Ugandan oil region weary of future". Retrieved from http://www.newvision.co.ug/D/8/220/746983/Oil (Accessed Dec, 2012)

The New Vision, Jan 23rd 2011, 'Uganda opts for local content to avert oil curse'. Retrieved from http://www.newvision.co.ug/D/8/220/744668/Oil (Accessed Dec, 2012)

The New Vision, Feb 10th 2011, 'Oil deals should be open, says law journal'. Retrieved from http://www.newvision.co.ug/D/8/220/746270/Oil (Accessed Dec, 2012)

The New Vision. Jan 13th 2011, 'Oil stirs up business in Hoima'. Retrieved from http://www.newvision.co.ug/D/8/220/743625/Oil (Accessed Dec, 2012)

The New Vision, Feb 24th 2011,'Inflation could hit 7% by June'. Retrieved from http://www.newvision.co.ug/D/8/220/747472/Oil (Accessed Dec, 2012)

The New Vision, Dec, 20th 2010: 'Oil a blessing to Uganda Kutesa' Retrieved from http://www.newvision.co.ug/D/8/19/741737/Oil (Accessed Dec, 2012)

The New Vision, March 11th 2011, 'Will locals turn into oil wealth?' Retrieved from http://www.newvision.co.ug/D/8/220/748710/Oil (Accessed Dec, 2012)

The New Vision, November 15th 2010: "Oil money to drive growth 10 percent. Retrieved from http://www.newvision.co.ug/D/8/220/738266/Oil (Accessed Dec, 2012)

The New Vision, Nov 4th 2010; 'Oil money to be shared with communities, Lokeris'. Retrieved from http://www.newvision.co.ug/D/8/220/737095/Oil (Accessed Dec, 2012)

The New Vision, Jan 19th 2011; 'Election cash could ignite inflation'. Retrieved from http://www.newvision.co.ug/D/8/220/744142/Oil (Accessed Dec, 2012)

The New Vision, Dec 31st 2010; 'Oil to create 120,000 jobs' Retrieved from http://www.newvision.co.ug/D/8/12/742508/Oil (Accessed Dec, 2012)

The New Vision. Jan 17th 2011; 'Museveni assures Banyoro on oil'. Retrieved from http://www.newvision.co.ug/D/8/13/744068/Oil (Accessed Dec, 2012)

The New Vision. Feb 20th 2011; 'President Museveni's Challenges'. Retrieved from http://www.newvision.co.ug/D/8/13/747129/Oil (Accessed Dec, 2012)

The New Vision, Feb 13th 2011; 'IMF, Uganda disagree over new policies' Retrieved from; http://www.newvision.co.ug/D/8/220/746558/Oil (Accessed Dec, 2012)

The New Vision, Nov 9th 2012; 'Uganda: MPs Disagree On Oil Penalties' Retrieved from; http://www.newvision.co.ug/news/637151-mps-disagree-on-oil-penalties. html

The Observer 2012: Uganda: Most Corrupt in East Africa; Retrieved from http://www. observer.ug/index.php?option=com_content&view=article&id=20686:uganda-most-corrupt-in-east-africa (Accessed Jan, 2013)

The Observer 2012: Oil Bill; Kadaga Flees House, Retrieved from; http://www. observer.ug/index.php?option=com_content&view=article&id=22350:oil-bill-kadaga-flees-house-

The Wall Street Journal, Friday 22nd Feb. 2013; 'Uganda Parliament Passes Second Oil Bill' Retrieved from: http://online.wsj.com/article/BT-CO-20130222-704832.html

Think Africa Press (2013) 1st Feb Issue: Ruth Peluse: Uganda's "Bad Politics"; Museveni, the Military and an Assertive Parliament; Retrieved from; http:// thinkafricapress.com/uganda/coup-talk-sign-weakened-regime-or-ruthless-bottom-line-nrm-museveni (Accessed Jan, 2013)

Transparency International (2005) *Avoiding the Resource Curse: What can we learn from the Case of Botswana?* EADI Conference. http//eadi.org/gc2005/confweb/papers/ Peter_Eigen.pdf (Accessed Jan, 2013)

Tripp, Ali Marri (2010). Museveni's Uganda: Paradoxes of Power in Hybrid Regime. Boulder/London: Lynne, Reinner Publishers.

Woo-Cuming, Meredith (ed.) (1999) The Developmental State, New York: Cornell University Press.

World Bank (2010): Africa's Pulse: An Analysis of Issues Shaping Africa's Economic Future. Africa's Pulse Series, 2.

CURRENT AFRICAN ISSUES PUBLISHED BY THE INSTITUTE

Recent issues in the series are available electronically
for download free of charge www.nai.uu.se

1981

1. *South Africa, the West and the Frontline States. Report from a Seminar.*

2. Maja Naur, *Social and Organisational Change in Libya.*

3. *Peasants and Agricultural Production in Africa. A Nordic Research Seminar. Follow-up Reports and Discussions.*

1985

4. Ray Bush & S. Kibble, *Destabilisation in Southern Africa, an Overview.*

5. Bertil Egerö, *Mozambique and the Southern African Struggle for Liberation.*

1986

6. Carol B.Thompson, *Regional Economic Polic under Crisis Condition. Southern African Development.*

1989

7. Inge Tvedten, *The War in Angola, Internal Conditions for Peace and Recovery.*

8. Patrick Wilmot, *Nigeria's Southern Africa Policy 1960–1988.*

1990

9. Jonathan Baker, *Perestroika for Ethiopia: In Search of the End of the Rainbow?*

10. Horace Campbell, *The Siege of Cuito Cuanavale.*

1991

11. Maria Bongartz, *The Civil War in Somalia. Its genesis and dynamics.*

12. Shadrack B.O. Gutto, *Human and People's Rights in Africa. Myths, Realities and Prospects.*

13. Said Chikhi, Algeria. *From Mass Rebellion to Workers' Protest.*

14. Bertil Odén, *Namibia's Economic Links to South Africa.*

1992

15. Cervenka Zdenek, *African National Congress Meets Eastern Europe. A Dialogue on Common Experiences.*

1993

16. Diallo Garba, *Mauritania–The Other Apartheid?*

1994

17. Zdenek Cervenka and Colin Legum, *Can National Dialogue Break the Power of Terror in Burundi?*

18. Erik Nordberg and Uno Winblad, *Urban Environmental Health and Hygiene in Sub-Saharan Africa.*

1996

19. Chris Dunton and Mai Palmberg, *Human Rights and Homosexuality in Southern Africa.*

1998

20. Georges Nzongola-Ntalaja, *From Zaire to the Democratic Republic of the Congo.*

1999

21. Filip Reyntjens, *Talking or Fighting? Political Evolution in Rwanda and Burundi, 1998–1999.*

22. Herbert Weiss, *War and Peace in the Democratic Republic of the Congo.*

2000

23. Filip Reyntjens, *Small States in an Unstable Region – Rwanda and Burundi, 1999–2000.*

2001

24. Filip Reyntjens, *Again at the Crossroads: Rwanda and Burundi, 2000–2001.*

25. Henning Melber, *The New African Initiative and the African Union. A Preliminary Assessment and Documentation.*

2003

26. Dahilon Yassin Mohamoda, *Nile Basin Cooperation. A Review of the Literature.*

2004

27. Henning Melber (ed.), *Media, Public Discourse and Political Contestation in Zimbabwe.*

28. Georges Nzongola-Ntalaja, *From Zaire to the Democratic Republic of the Congo.* (Second and Revised Edition)

2005

29. Henning Melber (ed.), *Trade, Development, Cooperation – What Future for Africa?*

30. Kaniye S.A. Ebeku, *The Succession of Faure Gnassingbe to the Togolese Presidency – An International Law Perspective.*

31. J.V. Lazarus, C. Christiansen, L. Rosendal Østergaard, L.A. Richey, Models for Life – Advancing antiretroviral therapy in sub-Saharan Africa.

2006

32. Charles Manga Fombad & Zein Kebonang, *AU, NEPAD and the APRM – Democratisation Efforts Explored.* (Ed. H. Melber.)

33. P.P. Leite, C. Olsson, M. Schöldtz, T. Shelley, P. Wrange, H. Corell and K. Scheele, *The Western Sahara Conflict – The Role of Natural Resources in Decolonization.* (Ed. Claes Olsson)

2007

34. Jassey, Katja and Stella Nyanzi, *How to Be a "Proper" Woman in the Times of HIV and AIDS.*

35. M. Lee, H. Melber, S. Naidu and I. Taylor, *China in Africa.* (Compiled by Henning Melber)

36. Nathaniel King, *Conflict as Integration. Youth Aspiration to Personhood in the Teleology of Sierra Leone's 'Senseless War'.*

2008

37. Aderanti Adepoju, *Migration in sub-Saharan Africa.*

38. Bo Malmberg, *Demography and the development potential of sub-Saharan Africa.*

39. Johan Holmberg, *Natural resources in sub-Saharan Africa: Assets and vulnerabilities.*

40. Arne Bigsten and Dick Durevall, *The African economy and its role in the world economy.*

41. Fantu Cheru, *Africa's development in the 21st century: Reshaping the research agenda.*

2009

42. Dan Kuwali, *Persuasive Prevention. Towards a Principle for Implementing Article 4(h) and R2P by the African Union.*

43. Daniel Volman, *China, India, Russia and the United States. The Scramble for African Oil and the Militarization of the Continent.*

2010

44. Mats Hårsmar, *Understanding Poverty in Africa? A Navigation through Disputed Concepts, Data and Terrains.*

2011

45. Sam Maghimbi, Razack B. Lokina and Mathew A. Senga, *The Agrarian Question in Tanzania? A State of the Art Paper.*

46. William Minter, *African Migration, Global Inequalities, and Human Rights. Connecting the Dots.*

47. Musa Abutudu and Dauda Garuba, *Natural Resource Governance and EITI Implementation in Nigeria.*

48. Ilda Lindell, *Transnational Activism Networks and Gendered Gatekeeping. Negotiating Gender in an African Association of Informal Workers.*

2012

49. Terje Oestigaard, *Water Scarcity and Food Security along the Nile. Politics, population increase and climate change.*

50. David Ross Olanya, *From Global Land Grabbing for Biofuels to Acquisitions of AfricanWater for Commercial Agriculture.*

2013

51. Gessesse Dessie, *Favouring a Demonised Plant. Khat and Ethiopian smallholder enterprise.*

52. Boima Tucker, *Musical Violence. Gangsta Rap and Politics in Sierra Leone.*

53. David Nilsson, *Sweden-Norway at the Berlin Conference 1884–85. History, national identity-making and Sweden's relations with Africa.*

54. Pamela K. Mbabazi, *The Oil Industry in Uganda; A Blessing in Disguise or an all Too Familiar Curse? Paper presented at the Claude Ake Memorial Lecture.*

CLAUDE AKE MEMORIAL PAPER SERIES

Published by Uppsala University and the Department of Peace and Conflict Research
Recent issues in the series are available electronically for download free of charge
www.pcr.uu.se/research/publications/CAMP/

1. Jinadu, L. Adele : Explaining & Managing Ethnic Conflict in Africa: Towards a Cultural Theory of Democracy (2007).

2. Obi, Cyril I: No Choice, But Democracy: Prising the People out of Politics in Africa? (2008).

3. Sesay, Amadu: The African Union: Forward March or About Face-Turn? (2008).

4. Boafo-Arthur, Kwame: Democracy and Stability in West Africa: The Ghanaian Experience (2008).

5. Villa-Vicencio, Charles: Where the Old Meets the New: Transitional Justice, Peacebuilding and Traditional Reconciliation Practices in Africa (2009).

6. Mohamed, Adam Azzain: Evaluating the Darfur Peace Agreement: A Call for an Alternative Approach to Crisis Management (2009).

7. Mbabazi, Pamela K: The Oil Industry in Uganda; A Blessing in Disguise or an all Too Familiar Curse? (2013)